THE WAY TO IMMORTALITY PUBLICIZED

Mark C. Wang

THE WAY TO IMMORTALITY PUBLICIZED

ISBN 9798992861709

To the Creator

Thank you

For everything

PREFACE

After reading this book, some readers will think it is the most important book in their lives and will follow the clues provided within to find the secret of immortality, other readers will think it is the most incredibly ridiculous book. Some other people will even deliberately belittle this book or attack the author after reading it. No matter which type of reader you are, please read this book with an open heart, open mind, and no ego. Because this book does not have gorgeous rhetoric and fancy writing, it only leads you to the truth of eternal life with facts, strong evidence, and rigorous scientific experiments.

TABLE OF CONTENTS

BACKGROUND

Can Our Limited Human Brains Really Understand This Infinite Universe?

For a long time, people thought that superpowers were a kind of pseudoscience and those who believed in superpowers were ignorant. However, the topic has been secretly studied by the world's top powers at a huge cost. The people involved in the research are the world's top physicists, psychologists, sociologists, and even philosophers.

In 2000, a declassified CIA file, "Project Star Gate" admitted the United States had conducted a secret project to study supernatural powers and, at the time discovered the existence of superpowers. The Project Star Gate also allowed the CIA to see the relics on the moon and Jupiter through the remote viewing function of superpowers or clairvoyance.

Project Star Gate was secretly carried out by the CIA and was not made public, so ordinary Americans had no evidence of superpowers. Unexpectedly, on the other side of the Pacific Ocean, major newspapers in China were causing a wave of enthusiasm with reports of the appearance of many children with special abilities. These reports also attracted the attention of American scientists. In the early 1980s, several

American scientists flew to China to participate in a special study to find the children with special abilities. It was not until August 11, 2000, when the CIA declassified the file called "China's Psychic Children" which provided records of the special study.

On March 11, 1979, China's *Sichuan Daily* published a news report saying that a boy named Tang Yu in Dazu County, Sichuan, could read with his ears.

Tang Yu was a fifth-grade student at the time. He and his classmate Chen Xiaoming were walking on the road when Tang Yu's ear accidentally touched Chen Xiaoming's shirt pocket while horsing around. The two Chinese characters FLYING GEESE suddenly appeared in Tang Yu's brain. It turned out that there was a pack of cigarettes brand-named "FLYING GEESE" in Chen Xiaoming's shirt pocket. Tang Yu's superpower was confirmed by the schoolteachers and then spread to the ears of the leaders of the Sichuan Provincial Party Committee. After a series of official tests, it was finally confirmed that Tang Yu's superpower of reading with his ears was real.

At that time, all the major newspapers in China reprinted this news, but the story did not end there. More children with superpowers like Tang Yu's were discovered, many of them could not only read with their ears but also with their noses, fingers, toes, and even armpits.

According to statistics from the Chinese National Science and Technology Commission, between March 14 and May 5, 1979, 12 provinces in China reported 20 people with special abilities, more than half of whom were children aged 9 to 15. At that time, the whole of China seemed to have set off a wave of special abilities. Even the April 1980 issue of *Nature Magazine* used the experiment of children with special abilities as the cover and the theme of the report. As time went by, this wave of enthusiasm did not subside, but spread to Hong Kong, Taiwan, Japan, and even attracted the attention of Western countries.

Marcello Truzzi, an American sociologist, founded the Center for the Study of Scientific Anomalies. He specializes in studying all mysterious phenomena and pseudoscience. When he saw a *Los Angeles Times* report stating there were many children with special abilities in China, he was immediately attracted. Truzzi began to learn about Chinese history, and he found that before communism, there were quite a few cases and stories of such human special abilities recorded in Chinese history.

Professor Truzzi learned that Qian Xuesen, the leader of the Chinese scientific community at the time and the father of China's atomic bomb, hydrogen bomb, satellite, and intercontinental missile, was the main promoter of human body science research in China at the time. Qian Xuesen strongly suggested to the then-Chinese Communist Party (CCP) government to study human special functions, which won the strong support of Zhao Ziyang, the top leader of the CCP, and Hu Yaobang, the second-in-command of the CCP.

Truzzi, Stanley Krippner-professor of mythology and parapsychology, Harold E. Puthoff- participant of the Stargate Project and parapsychologist, and two other medical experts formed an American expert group. In this way, a perfect elite team consisting of three famous American experts in parapsychology and pseudoscience and two medical experts arrived in Beijing on October 18, 1980.

The Beijing experts who met with the American expert group introduced the relationship between human special functions, qi and qigong. The Beijing experts also shared the effect qigong cures has on curing diseases. According to research by the CCP government, more than 20% of 3,100 patients with chronic diseases were completely cured after practicing qigong for 3 to 5 years. More than 40% of the patients had obvious improvement in their conditions. 20% of the patients had a general improvement in their conditions.

So, what is Qigong? You can think of Qigong as a simple version of yoga that anyone can practice. It mainly consists of hand movements and meditation. The most powerful Qigong in the world, which can cure chronic diseases at a rate of 98%, is the famous Falun Gong.

After three days of academic discussion, the American experts finally met the Chinese children with special abilities. An experiment was conducted in the hotel lobby. More than 50 researchers observed the test. Four little girls had to complete test items given by Chinese and American experts. The Chinese experts brought folded papers with words written on them. The American experts brought sealed bottles that were prepared in the United States. The experiment was simple: the experts asked the subjects to say what was on the paper and in the bottle.

Although the American team hoped that the experiment could be conducted more rigorously, according to the CIA report, Mar, the translator that day, spoke ambiguously in Chinese. The American team even doubted whether he had told the children the answer. Perhaps because of the language communication problem, the experiment ended in chaos. As a result, everyone moved to the conference room to start a formal experiment. The result of the experiment was that the children did not guess the content of any of the slips correctly.

Then, the American experts went to their second stop in Xi'an, China, where there were three children with special abilities like those in Beijing. The experiments resulted in none of the three children being able to guess the contents of the notes.

The American team went to Shanghai, their last stop, where trouble arose. The team arranged to meet with people from Nature magazine and doctoral students from Shanghai University who were interested in human supernormal abilities. When the American experts arrived at the hotel, they found that all the meetings were canceled. When they were in Xi'an, Yu Guangyuan, then deputy director of the State Science and Technology Commission of the CCP, criticized the research and promotion of human supernormal abilities at a scientific report meeting held by the Academy of Sciences. This senior official of the CCP who firmly believed in Marxist materialism claimed that the research on human supernormal abilities was nonsense and superstition. He also began an investigation committee to expose falsehoods about human supernormal abilities.

Due to the Chinese version of cancel culture, the shadow of the Cultural Revolution was still vivid in people's minds. Most scientists who studied human special abilities were forced to withdraw from the research. Fortunately, a journalist couple who worked for Nature magazine found the American team and told them about the special human functions they had experienced and seen in their journalistic careers, such as children who could emit infrared rays, unusual brain waves, and magnetic field signals; a child who saw a pregnant woman with a deformed fetus in her belly, which late, was confirmed with an ultrasound scan.

Professor Truzzi's team heard many similar stories in China, but for them, scientific evidence was most important. Professor Truzzi's team conducted a third and final experiment with Chinese scientists. However, the Chinese scientists had a different understanding and definition of the rigor of scientific experiments from the American team, so the experiment did not convince the American team. This conclusion led Professor Truzzi to requesting to bring some children with special abilities to the United States to replicate the experiments under the rigorous experimental standards of the United States to prove whether special abilities exist or not. Truzzi was rejected by CCP officials and after returning to the United States, the American team had to give up the idea of replicating these experiments.

A Turning Point Emerged

In 1981, one year after Professor Truzzi returned from China, a young Taiwanese men whose name was Lee Si-chen returned to Taiwan with his PhD in Electrical Engineering from Standford University. Dr. Lee Si-Chen used the strict American testing standards to complete the human special ability experiment that Professor Truzzi had been unable to complete in both Taiwan and the United States.

Dr. Lee Si-Chen became a professor of Electrical Engineering at National Taiwan University (NTU), and later the president of NTU, and completed the experiment that Isaac Newton could not complete with mathematics. He verified the existence of human special abilities through rigorous scientific experiments and proved the existence of God and other sacred higher life forms, as well as the dimension in which they exist. More importantly, these scientific experiments can repeatedly prove the existence of God and other divine beings.

Perhaps readers do not believe that a small country like Taiwan is capable of successfully conducting scientific experiments that can prove the existence of Divine beings like Jesus and Buddhas. I want to tell these readers that Taiwan is not the only country to use scientific experiments proving the existence of God.

Communist China has also used scientific experiments to prove the existence of God. They did not use one scientific method to like Taiwan. Multiple academic institutions in Communist China have used various scientific methods to prove the existence of divines.

Doesn't this sound a bit ironic? Communist China is an atheist country, but they have repeatedly used different scientific methods to prove the existence of divine beings.

What is even more shocking is that the original purpose of the human supernormal ability experiments conducted in Taiwan and Communist China was to verify the mysterious phenomena mentioned in a book called *Zhuan Falun* by Master Li Hongzhi. During these experiments, the existence of God was accidentally discovered.

The Book from Beyond Earth

What kind of book is *Zhuan Falun*? In the words of more than 100 million readers who have read *Zhuan Falun*, when you finish reading it, you will know the unimaginable magic and power of the book holds. This book explains the mysterious phenomena that humans cannot understand. For example, various superpowers of the human body, human body rejuvenation, the relationship between humans and gods, prehistoric civilizations, ancient prehistoric technology, parallel universes, reincarnation, Akashic records, aliens, the truth of the universe, supernatural methods of eliminating diseases, the method of immortality, etc.

Since the knowledge in *Zhuan Falun* is not considered human knowledge, it has attracted the attention of the scientific community in Taiwan and Communist China. After more than 30 years of research, these countries have used scientific methods to prove one by one that the mysterious phenomena mentioned in this book are true.

Zhuan Falun is also the main ideological guide for Falun Gong practitioners. As this book carries the powerful force of the highest principles of the universe, it changed the entire social landscape of communist China in the 1990s and made more and more Chinese people and social elites, including the army, police, and government officials in the communist regime, strictly require their own behavior according to Truthfulness, Compassion, and Forbearance. In addition, the memory of the Tiananmen Massacre and the Tank Man in 1989 was deeply engraved in the minds of the Chinese people. In such an environment, the CCP not only quickly lost the support of the people, but also gradually lost the support of many party members and senior officials. This directly threatened the CCP regime based on lies and violence.

As a result, Jiang Zemin, the top leader of the CCP at the time, used his absolute power as a dictator to disregard public opinion and even the opposition of the majority of party members and began an unprecedented 25-year-long brutal persecution and suppression of Falun Gong and banned *Zhuan Falun*.

Due to the charm of *Zhuan Falun*, many leading figures in China's scientific community are still studying the book's contents. At the same time, the CCP also wants to know the mysterious phenomena in *Zhuan Falun*, especially the special functions of the human body, for military purposes, so it has been providing government-funded research towards scientific institutions to support researchers to verify and develop the special functions of the human body mentioned in *Zhuan Falun*.

The purpose of writing this book is to tell you that if you want to understand the truth about the human body, life, the universe, and the divine world, want to change your destiny, and want to understand the true meaning of immortality, then the book *Zhuan Falun* is a must-read book.

Zhuan Falun is similar to *The Secret of the Golden Flower*, a best-selling book on immortality in Europe written by the great psychologist Carl Jung in the early last century. However, *Zhuan Falun* covers more truths about the universe. Moreover, *The Secret of the Golden Flower* was written based on Jung's limited personal understanding of the Chinese Taoist immortality book *Taiyi Jinhua Zongzhi* or 《太乙金華宗旨》. Therefore, *Zhuan Falun* can be a good supplement and explanation to *The Secret of the Golden Flower*.

CHAPTER 1:

Scientific Methods Have Proved the Existence of the Spirit World

I. The World's First Scientist to Unveil the Mysterious Veil of Different Dimensions

At present, there are at least 100 million people in the world who believe that the knowledge in the book of *Zhuan Falun* comes from outside the earth and the book itself has a miraculous healing effect. Because *Zhuan Falun* is a prehistoric science that teaches people to improve morality, eradicate diseases, release souls, and ultimately make people immortal, is there a third party who can use scientific experiments to prove it? Do the Buddhas, Taoists, and gods mentioned in *Zhuan Falun* exist? If they do, where are they?

The whole world knows Taiwan's research in semiconductor technology is at the forefront of the world. The chips produced by TSMC determine the future of our human technology. However, the world does not know Taiwan's research in the science of the spirit world is also at the forefront of the world. Dr.Lee Si-Chen is the leading figure in researching the science of the spirit world.

> *"Breaking through the boundaries of science and opening Pandora's box of truth about the universe—*

"Use science to prove the existence of the spiritual world!"

It turns out that there are really gods, Buddhas, and spiritual realms outside our world?

The real universe is not just a single space-time Yang world, but a plurality of time-spaces with both Yin and Yang."

"Lee Si-chen, PhD in Electrical Engineering from Stanford University and former president of National Taiwan University.

Based on the belief that "experiment is the only criterion for testing truth", 25 years of hard work have been devoted to scientific empirical research on special functions such as finger reading and telekinesis.

Witnessed by many physicists and psychologists, found the use of religious sacred words such as "God", "Buddha", "Bodhisattva", "Jesus", and etc.

It allows high-functioning people to "see" visions on the brain screen and can communicate with highly intelligent beings in the spiritual world through consciousness, Gradually unraveling the mysteries of contemporary science—Dark matter, dark energy, super-light quantum entanglement, special functions, alien civilization!"

The above is the promotional slogan printed on the back of the book *Science of the Spiritual World.*

The Science of the Spiritual World by Dr. Lee Si-Chen has not yet been translated into English and other languages.

Dr.Lee Si-Chen graduated from the Department of Electrical Engineering at National Taiwan University in 1974 and received a Ph.D. in Electrical Engineering from Stanford University in 1981. He has been a professor at the Department of Electrical Engineering at National Taiwan University since 1985. His expertise is in the research of semiconductor optoelectronic materials and devices, nanostructure optoelectronic devices, qigong, special functions, and information fields. He has served as the head of the Department of Electrical Engineering at National Taiwan University, the Dean of Academic Affairs at National Taiwan University, the President of National Taiwan University (2005-2013), the Chairman of the Chinese Society of Electrical Engineering, the Chairman of the Association of National Universities and Colleges, and was elected as a Fellow of the Asia-Pacific Academy of Materials (1997) and a Fellow of the International Institute of Electrical and Electronics Engineers (IEEE) (2002). His book, "Physics of Semiconductor Devices" won the Golden Tripod Award in 1995, and the edge-thinning structure of heterojunction transistors he invented in 1985 has become an industrial standard. Dr.Lee Si-Chen retired from National Taiwan University in October 2018.

From the above resume of Dr.Lee Si-Chen, we can see that he is a no-nonsense scientific guy.

One word opens the door to the spirit world.

On August 26, 1999, a strange phenomenon occurred during Dr. Lee Si-Chen's finger reading experiment. Witnessed by more than a dozen physicists and psychologists, Dr. Lee opened the door to another world with his finger reading of the word "Buddha."

The little girl in the experiment could see the correct, static words when she looked at other words. But when she saw the word "Buddha," she said she saw a glowing person smiling at her. They tested several more "Buddha" words in a row, and the child said she saw a glowing screen, heard laughter, and saw a glowing person running around in the laboratory, but she couldn't see the words.

Next, they asked the child to look at the words "Bodhisattva" and "Jesus." The result was either a glowing person or a cross.

Could this be an illusion? Dr. Lee says, "NO" because similar results were obtained from other children's experiments. This is a scientific experiment that can be repeated. Among them, a little girl described what she saw: "There is a temple, and there is a person at the door, flashing."

Another boy described: "There was a bald man holding a rosary."

Xu, another little girl, touched the words "Buddha" and "Jesus", but she only saw the words, not the images. However, when she touched the words "Maitreya Buddha", she saw the image of Maitreya Buddha. When she looked at "Medicine Buddha", she saw a bright light like the sun, which made her eyes unable to open.

What would Sun Chulin, a lady with special abilities and works at the Institute of Human Body Research at Wuhan University of Geosciences in China, see? When Mrs. Sun touched the word "Buddha", she saw thousands of golden lights.

The Internet of the Spirit World

On February 24, 2024, at the annual meeting of the Chinese Association of Scientists and Engineers in San Diego, California, when Dr. Lee gave a speech on the theme of "Exploring the Mysterious Power of the Aura (Tortion Field) with Science." He also discussed why everyone sees something different Dr. Lee started to figure it out. One day he suddenly had an inspiration and said that if the information field is the Internet, the sacred words are the URLs in it, which can guide people into higher worlds. Some people have low functions, perhaps like a low-end version of the browser, and many websites cannot be browsed. People with high functions have powerful browsers and can browse higher and more extraordinary scenes.

Dr.Lee Si-Chen elaborated that in the information world, people are computers, the information field is the Internet, and the sacred words are websites. Various sacred figures build their websites on URLs. People with special abilities, whether they can read with their fingers or have opened their third eye, can browse these websites and communicate with the gods.

Based on this concept, Dr. Lee started to study and conduct more experiments. As a result, he found these URLs seemed real, and it was not okay to enter them incorrectly. For example, if the word "Buddha" were slightly changed, the strange phenomenon would disappear. The same was true for other languages. When testing the English word "Buddha", he accidentally missed a letter, and Miss T, who was being tested, could not see the phenomenon but could only see the letters. In addition, the names of the gods must be written neatly. Once, an experimenter wrote the words "Mazu", a local deity in Taiwan, in different sizes. Miss T stated the image of Mazu became blurred and distant. Dr. Lee later said with emotion: "To communicate with heaven, one must be sincere and upright and write the sacred words in a proper manner to ensure safety."

At this point, some people may be curious. Mainland China uses simplified Chinese characters. Many of the characters are missing parts compared with the traditional Chinese characters that Dr. Lee Si-Chen tested. Can the gods recognize them? Dr. Lee tested simplified Chinese characters. He tested the two characters "Guanyin/ 观音"(a female Chinese Bodhisattva), in which the character "Guan/ 观" was simplified. What was the result?

The traditional Chinese characters for "Guanyin" can make people see a bright light, but no words. After changing to simplified Chinese characters, the brightness is greatly reduced, and the words appear in the background. Dr. Lee Si-Chen explained the sacred meaning of these two characters has been greatly weakened. For two thousand years, Guanyin and the Chinese people have been communicating with traditional Chinese characters. The sudden creation of simplified Chinese characters seventy years ago may have caused a poor connection. Maybe it will take a few hundred years for the Chinese people to re-establish a stable communication channel with the Guanyin Bodhisattva. But in those few hundred years, will the Bodhisattva continue to bless the people who use simplified Chinese characters?

Anyone who has learned Mandarin Chinese knows that there are two writing systems for Chinese. The first is called Traditional Chinese. This was used in China for thousands of years before the advent of Communism. Taiwan and Hong Kong still use this writing system. The second is called Simplified

Chinese, which is the writing form used in China today. It is a simplified version of the traditional Chinese characters. Simplified Chinese characters are a product of the Chinese version of the Cancel Culture movement, also known as the Cultural Revolution, launched by the CCP in the 1960s. The real purpose of adopting the simplified characters is to destroy the traditional culture and beliefs of the Chinese people to hinder people's communication with divine beings and ultimately keep people away from divine beings.

At this point, you may ask, do gods, Buddhas, and Jesus exist in this world? Dr. Lee said, yes. Based on the experimental results, not only the founders of major religions have established websites in the information field, but also the saints, immortals, and gods worshipped by the people in history seem to have websites in the information field. They are real existences, but they exist in another space. Science will guide us into their world. I hope one day we will greet and say "Hello" to them directly on the screen.

It's quite interesting, right? I wonder if one day we will encounter gods in the world of the Internet.

Dr. Lee Si-Chen said that he had practiced qigong for a long time and had a deep feeling for it. "I knew this thing was real," he said. "China has been talking about qigong for thousands of years, and hundreds of millions of people are doing this. You can't fool people. So, I plunged into the scientific research of qigong." (Dr. Lee Si-Chen is not a Falun Gong practitioner.)

Later, special abilities such as finger reading aroused his interest. He began to design experiments and repeatedly proved the existence of divine beings. Why can fingers read without light? Can all words be recognized equally? The challenge of his peers inadvertently gave him a "light bulb" and he began to turn to the study of the aura (torsion field) and how it communicates. After more than 30 years of in-depth research, he explained that "the spirit world is around us" from the perspectives of physics, complex functions, quantum mechanics, etc. He tried to provide a scientific basis for the mysterious phenomena described by the Chinese Taoist "Qi Men Dun Jia/奇門遁甲" and the American inventor Nikola Tesla.

Talking about the torsion field and his related research being questioned by some scientists or called "pseudoscience", Dr.Lee Si-Chen said there are many things in the universe that people have not yet understood.

He said that a major discovery at the end of the 20th century was the discovery of dark energy in the universe. Dark energy accounts for 73% of the energy distribution in the universe, while another type of energy called dark matter accounts for 23%; the remaining 4% is normal.

He explained the normal part, the world that people can know and understand normal matter, only accounts for 4% of the universe. "And dark matter and dark energy, people still don't know what they are. In other words, you don't know 96% of the world. You live in a 4% world, called the world of materialism. We are trapped in this 4% world under the philosophical foundation of materialism." He laughed, "You study the '4%' world, I study the '96%', what right do you have to criticize me?!"

Dr. Lee Si-Chen hopes the speech can open people's eyes. "Believe that these things are scientific. Whether it is qigong, special abilities, or finger reading, they can be verified in the laboratory. They are facts.

Don't think they are supernatural things. They are not. They can be verified through experiments." He also emphasized, "Qigong is believed by most Chinese people because it is a Chinese tradition for thousands of years. It is impossible for it to deceive people for thousands of years and hundreds of millions of people. Experiments repeated in different laboratories, psychological changes, physiological changes, these are all scientific facts."

Conducting these studies has sent shockwaves through the academic world. In the face of doubts, Dr. Lee Si-Chen explained that science is about researching things that are not understood. Scientists who question his research are against science. He explained that we try to understand things that we do not understand. "As a result, they criticized me. I only accept one thing, if there is a problem with the experiment I did, and the procedure is not rigorous enough, then I accept their criticism. Experiments are the only standard for testing the truth!" He continued to emphasize that scientists who opposed him did not do such experiments, so how do they know that the spirit world does not exist? "Then aren't they superstitious? They are scientific superstitions!"

Dr. Lee pointed out that these people always think that things not in the scientific framework are pseudoscience and should not be studied. What they say is contrary to the spirit of science. The spirit of science is that if I don't know, I must find out its principles and its basis. In this way, won't the framework of science become larger?

He proved the existence of God for Newton.

I believe that many science fiction fans have read Liu Cixin's novel *Three-Body Problem*. The movie adapted from this novel was released in 2024 and once again triggered a craze. The background of the *Three-Body Problem* is that the Trisolarans live in Alpha Centauri, which has three stars. The three stars cannot exist stably, resulting in the Stable Era and the Chaotic Era alternating..

The gravitational problem of three bodies in its traditional sense dates in substance from 1687, when Isaac Newton published his *Philosophiæ Naturalis Principia Mathematica.* Newton was trying to discover if any long-term stability was possible, especially the system of our Earth, the Moon, and the Sun.

Newton was thinking about the many celestial bodies in the solar system and how they interact with each other. Is the solar system a stable structure? Newton first assumed the solar system was stable because humans survived for so many years. However, Newton could not mathematically prove that the solar system was stable. He even suspected that if he solved it mathematically, he would find that the solar system was unstable. Newton's explanation for the reason why the solar system can be stable is that there is a first mover. This first mover pushes the solar system, and this first mover is God. Moreover, the solar system is a multi-body galaxy, which may be unstable and will become chaotic as it rotates. In the moment of chaos, God will take action to set it right and restore it to a stable state.

Newton wanted to use mathematics to prove when the solar system would be in chaos. By calculating when the solar system would be in chaos if it was not in chaos at that time, he would know that there was a supernatural force intervening at that time; this supernatural force was God. Newton would figure out when God would take action. In this way, he could use mathematics to prove the existence of God. But Newton died without completing this task.

Dr. Lee Si-Chen has not only proved the existence of God but also found where God is. Why has Dr. Lee not won the Nobel Prize? Why has his book not been translated into English or other languages?

He has done tremendous scientific research on psychic abilities, different dimensions, and the existence of Jesus, Buddhas, and other deities!

I would like to add that both the finger-reading experiment mentioned by Dr. Lee Si-Chen and the existence of different gods, Buddhas, and Taoists in different dimensions have already been mentioned and explained in *Zhuan Falun.*

Even the question that had been bothering Dr. Lee Si-Chen during the experiment–"Why does everyone see something different?" can be found in *Zhuan Falun.* Dr. Lee explained that some people have low functions, perhaps like a low-end browser, and cannot browse many websites. People with high functions have powerful browsers and can browse higher and more extraordinary scenes. But had Dr. Lee Si-Chen read *The Second Talk* in *Zhuan Falun,* he would have known that people who have cultivated to a certain level can only see scenes at that level. The person cannot see the truth beyond this level, nor does he or she believe it, so he or she only believes what was seen at his or her level is correct.

II. The CCP's Science of Human Special Abilities

Let me first introduce Qian Xuesen. Xuesen founded China's renowned 507th Institute of Human Sciences. He is also the father of China's atomic bomb, hydrogen bomb, and satellite as well as ICBM.

Western friends may not be familiar with him. This leader of the red China's scientific community was born in Shanghai in 1911. He was gifted since childhood and went to the United States to study rocket technology at 24. He was very talented in this area and published many important papers. He was hired as a full professor at MIT at a young age and was awarded the rank of Air Force Colonel by the U.S. Department of Defense.

However, after the Cold War began in 1949, the FBI found that he had participated in some social activities organized by the CCP. With the Korean War on the verge of breaking out, the National Security Agency began to get nervous.

In September 1950, Qian Xuesen was taken away by the U.S. Immigration and Naturalization Service. After being released on bail, he was questioned at hearings for several weeks. At that time, the prosecutor asked him: "Are you loyal to communist China?" Qian Xuesen said: "I am not." The prosecutor asked: "Who are you loyal to?" Qian Xuesen replied: "I am loyal to the Chinese people." Although Qian Xuesen stated his position, under the circumstances at the time, he still could not be trusted by the U.S. Department of Defense, and he could not continue to do his favorite rocket research.

In 1955, China invited Qian Xuesen back to China in exchange for Korean War prisoners of war. With the support of this rocket genius, China rapidly began to develop missiles, nuclear weapons, and artificial satellites. In 1999, Qian Xuesen was named "Two Bombs and One Satellite Hero" by the Chinese government, which can be regarded as a lifetime achievement award.

However, Qian Xuesen, who was nearly 80 years old, was no longer interested in rockets. He was more concerned about human science. He also proposed the concept of human science. He said: "Human science is a large department in the modern scientific and technological system. It is on an equal footing with natural science and social science." And "Qigong is the key to open the door to human science."

The Cultural Revolution, or the Chinese version of Cancel Culture, ended in the 1980's, and the country was beginning to open. Everyone was curious about new things, and strange stories often appeared in newspapers and magazines. The phenomenon of human supernormal abilities was the most widely spread and hotly discussed topic, such as reading with ears, taking objects from a distance, bending iron spoons with the mind, etc. At that time, there were even people who opened training courses to teach children to develop supernormal abilities. The streets, alleys, and riverside parks were dominated by qigong enthusiasts. Various qigong schools emerged, and qigong classes blossomed everywhere.

At this point, some people may ask: No matter how open-minded the Chinese government was in the 1980s, it was still an atheist government. Could it allow the qigong craze to flourish? You know what, Qian Xuesen played an indispensable role in this.

Qian Xuesen believed that Qigong was a science. To promote this emerging science, Xuesen wrote to the government repeatedly and promoted the concept of human science. It is said that Hu Yaobang, then General Secretary of the CCP, changed his opinion after reading Qian Xuesen's letter. In 1982, under the instruction of Hu Yaobang, the Central Propaganda Department issued a "three-no policy", which was to "not report, not debate, and not punish" Qigong and human science research. The policy allowed a small number of people to conduct research and allow academic exchanges. It was tacit approval.

From then on, the 507th Institute began to conduct experiments on special abilities and many of the experimental results were published in professional journals. However, because it was a long time ago and there was no Internet, we could not find the published papers. What we can find on the Internet is the book *On Human Science and Modern Technology*, which is a collection of speeches and reports made by Qian Xuesen at the 507 in the 1980s. These reports cover a range of aspects, including human science, system science, qigong, traditional Chinese medicine, special abilities, etc.

However, on the CIA website, the title of the book *On Human Science and Modern Technology* was translated by the CIA into "Chinese Journal of Somatic Science" and only 57 pages of the book's content were made public.

What a Chinese American Professor Observed at 507th Institute

Mr. Li Youfu (Doctor of Traditional Chinese Medicine, World Cup Wushu Champion, Chairman of Wushu Competition Judges) who currently lives in Los Angeles is a rare well-known figure in China who has both martial arts skills and qigong functions, as well as theoretical research skills. He was once known for his ability to diagnose patients remotely without contacting them. In 1987, he was hired as the Executive Deputy Director of the Beijing Institute of Traditional Medicine and served as an associate researcher at the Research Center of the Chinese Society of Human Sciences under the leadership of Qian Xuesen. From 2005 to date, he has been a professor at Alhambra Medical University in Los Angeles.

Because Li Youfu had the ability of remote diagnosis, he was invited to the 507th Institute to serve as an associate researcher at the Chinese Human Science Research Center under the leadership of Qian Xuesen. Li Youfu's ability of remote diagnosis was also unique. Sitting opposite a person, he could see where the person's illness was. However, he said that this ability was acquired through practice, and he could not be considered a great qigong master. He met a little girl at the 507th Institute who could use the ability of the third eye to diagnose illnesses.

The child could see cancer cells on the patient and said they were ugly. Li Youfu said, "That's strange. Cancer cells are so small. Which eye do you use to see them?" The child said that she could magnify the diseased part to the size of a house, and the cancer cells would be as big as a watermelon, and she could see them very clearly. Not only that, but she also knew the relationship between her parents in their previous lives. This ability is called "the power of knowing fate." The child said that her father saved her mother in her previous life, so they became husband and wife in this life.

Li Youfu also met a little boy who was less than 10 years old. He could see aliens. The boy said that there were big aliens and small aliens. The smallest one was very small. It could dance on his hand and communicate with him. They played together like friends. Then the boy said that the United States already had stealth aircraft that was built with the help of aliens. The child also drew a sketch. After a while, the photo of the F-117 Nighthawk bomber was published in the newspaper. When compared with the little boy's sketch, it was the same, with almost no difference.

The CCP's Experiment Broke Through the Parallel Universe

The New Space 1025 Laboratory was established by the CCP in Yunnan in 2011. This Laboratory is an experimental institution that combines philosophy, physics, and parapsychology. The main person in charge of the project is Professor Gong Zhebing of the Department of Philosophy at Wuhan University. They used the adapted Taoist path training method to train a group of children with superpowers. They completed experiments such as moving objects with thoughts and sending cameras to other spaces with thoughts. The children and adults with special abilities in the experiment teleported the miniature cameras into another time and space and then captured footage, proving the existence of parallel universes.

They have successfully conducted experiments on teleportation, which is to transport objects from our space into another space. When the object disappears out of thin air in our space, they then move the object back to our space from another space.

To the mainstream scientific community in the United States, teleportation is just a hypothetical idea, a common topic in science fiction and other popular culture. It is often associated with time travel.

I don't know what progress the American scientific community has made in studying teleportation, but I know that teleportation is no longer just a topic of science fiction for their Chinese counterparts.

Professor Gong Zhebing had the idea of sending a GPS terminal tracker with real-time satellite tracking and positioning function into another time and space, hoping to track the trajectory of an object in another parallel time and space after it disappears.

They put the GPS tracker into a metal box. The person with special abilities teleported the GPS tracker to another time and space, but the box was not opened. Then the software connected to the satellite on the experimenter's smartphone and displayed the starting point, end point, speed, longitude, latitude, and the entire running track of the GPS terminal tracker.

Most importantly, similar experiments could be done repeatedly with consistent experimental results.

The results are convincing because they conducted the experiment in China and the GPS data is stored in the United States. They can't come to the United States to modify the data and patterns or make a false track.

Later, they continued to improve the experiment, adding a micro-camera and recorder to the GPS locator to teleport into another time and space, and recorded the images in another time and space. But they found that the video was completely black and there was nothing.

They added a miniature flashlight to the device and teleported it into another dimension. This time the camera successfully captured the scene; however, most of the footage and sound were strangely washed out.

A total of 57 minutes of footage was taken of the object teleported into another time and space. However, only three minutes of footage was saved, and it was blurry. The rest of the footage was inexplicably erased by an unknown force.

In the three-minute-long footage that has been preserved, a piece of paper with «Max», «Address: Jinsha Town 503" written on it, the image of a pen, and a piece of bread wrapped in plastic were captured. There was also a strange sound.

The superpower kid who sent the camera into another dimension was named Zhang. Zhang used his third eye to see the scene in another dimension. He explained that there was another identical person in another dimension, but his name was Max. Max lived in another dimension, at No. 503 in a villa area called Jinsha Town.

All the operations in the experiment were performed by Zhang in another space, that is, Max. Zhang himself watched himself operate in another space and time . But in reality, Zhang was talking to the staff in the laboratory.

To solve the problem of not being able to shoot the images in another time and space, they later improved and teleported a smartphone with infrared photography function to shoot, but when the phone returned, the memory card that saved the video was inexplicably missing. It could not be found, and no one knew what happened. The 1025 laboratory kept the data and results of this experiment confidential and did not announce it to the public. It seems that a higher-level life is watching all this.

Master Li Hongzhi has explained the knowledge of parallel universes many times. He explained the relationship between parallel universes and dreams in this way:

> *Let's take up the topic of dreams. Spiritual teachers are reluctant to try to explain how they work. These are exceptionally complex phenomena, for at the time of your birth other parallel versions of you were born simultaneously in multiple dimensions of the universe, and together they form a complete*

> *body with you, linked via the mind; you also have inside your body a true soul, a secondary soul, and a wide array of supernatural beings, each with its own appearance; and every cell, along with your internal organs, reflects your appearance and energies in other dimensions... (Zhuan Falun, p371, Li Hongzhi, 1994)*

While hundreds of millions of Chinese people have improved their moral character, cured their diseases, gained supernatural powers, and learned the truth about the universe by practicing *Zhuan Falun*, the CCP declared Falun Dafa a "cult" and launched a brutal persecution against Falun Gong for more than 30 years. At the same time, this shameless communist regime is still studying almost all the supernatural powers mentioned in *Zhuan Falun.*

III. The CCP's Universities Studying the Contents Mentioned in *Zhuan Falun*

There is a well-known materialist atheist and communist in China–Sima Nan. Sima Nan was once famous for exposing fake qigong. Many "qigong masters" of the time were disciplined by him. However, this atheist who likes to define people with special abilities as liars and special ability phenomena as superstition never dares to provoke the special ability lady named Sun Chulin. This is because the human science research that Ms. Sun participated in was a government action funded by the CCP for more than 30 years. Sima Nan only dares to deny folk qigong masters like Li Hongzhi, the founder of Falun Gong, because he knows what it means to deny the government's behavior in a dictatorial country.

Ms. Sun Chulin, a researcher at the Institute of Human Science of Wuhan University of Geosciences, has had supernatural powers since childhood. She was born in Wuhan in 1957 and was sent to her grandmother's home in Nanjing to be raised when she was 10. By chance, when Chulin was helping her grandmother with farm work, she saw a jar filled with copper coins buried underground. This was the first time she discovered she had supernatural powers. However, in the general environment at the time, she did not make it public.

In 1979, Tang Yu, a 12-year-old primary school student in Sichuan Province, China, sparked the qigong craze across the country when it was discovered he had the ability to read with his ears,.Sun Chulin, who had returned to Wuhan to work as a kindergarten teacher, tried it at home and found that she could read with her ears, and it was not difficult for her to see through the human body. At that time, special powers had been generally accepted by everyone, and Sun Chulin generously shared her superpowers with the outside world. Soon, she was transferred to work at the General Hospital of the People's Liberation Army of China, where she used her fluoroscopy ability to see the location of lesions for patients.

In 1987, Sun Chulin met Professor Shen Jinchuan of the Institute of Human Sciences at Wuhan University of Geoscience, and she began 18 years of research on special functions as an assistant researcher. Her superpowers have developed more and more over the past 18 years, finally reaching 60 types, and each function challenges the existing knowledge of physics and biology.

In April 1996, at the "Special Phenomenon Physics Seminar" held in Tanzhe Temple near Beijing, Sun Chulin completed her three superpower experiments under the supervision of many scientists. This seminar was hosted by the Chinese Academy of Human Sciences led by Qian Xuesen. At that time, almost all scientists engaged in human science research on the mainland were present. Dr.Lee Si-Chen, the leader of Taiwan's superpower research community, was also invited to participate. Later, he wrote Chulin's three superpower experiments into a book he and his wife co-authored, called *"Unbelievable–Scientists Explore the Mysterious Information Field"*.

Teachers from other dimensions

Interestingly, Sun Chulin said she learned many of her superpowers from masters in another dimension, one of whom was the image of Laozi, the founder of Taoism. Her special abilities may include Taoist kung fu. The teachers in another dimension taught her kung fu in a very strange way. They all came to her when she was meditating and practicing at night, just like watching a movie, and appeared on the screen in her brain. Since only she could see it, she initially suspected that it was an illusion or a dream.

However, one thing happened that made her believe that a master was teaching her. That day, she heard that a well-known Chinese person with special abilities could shake pills out of a sealed medicine bottle, and she wondered if she could do it. That evening, a master with a whisk came and said he would teach her this skill. With a wave of the whisk, she had a small medicine bottle in her hand. Then the master taught her to soften the glass bottle and gently shake it. The pills in the bottle fell out and scattered all over the floor. When Sun Chulin woke up in the morning, she was still not sure whether it was true or not. When Professor Shen heard about it, he said, "Why don't we try it, and we will know?" So, he picked up a bottle at random, threw a few coins into it, and asked Sun Chulin to shake it. Sun Chulin did the same thing and shook it gently as the teacher taught her; unexpectedly, the coins fell out.

Sun Chulin believed that the teachers existed. Although they could not be seen with the naked eye, she believed these masters came from the "high intelligence information field," which is the high-level space. There are six or seven masters, one being an alien. The first time she saw the alien, she was meditating on a mountain near her home. When she walked down the mountain, an alien in a golden space suit followed her. He looked like Uncle Martin from Mars, with an antenna on his head and a metal rod in his hand, like a magic wand. If you want something, point it in the air and it will come.

In addition to teaching her skills, this alien teacher also took her to travel and visit other planets, such as Mars. Sun Chulin said there is water on Mars, and it was originally inhabited by people. It is like the story told by the Russian Martian boy circulating on the Internet. Many years ago, the Martians and aliens drove their respective spacecraft and fought fiercely, destroying everything on the ground, so now the surface of Mars is desolate. Some beings live underground, but they are no longer the natives of Mars, so the red planet has been occupied by aliens.

Mind drilling

The second experiment was to drill a hole with the mind. The experimental object was a ten-yuan coin from Taiwan provided by Dr. Lee. This kind of coin is rare in China, greatly reducing the possibility of cheating. Sun Chulin held the coin in her hand. After about 10 minutes, she opened her fist and a small hole with a diameter of about one millimeter appeared in the coin. How did this hole get made in full view of the public?

Mrs. Sun Chulin shared with everyone:

> *"First, you need to empty your mind and be quiet. It feels like you are in another time and space. Then a screen will appear in front of you. On the screen, there is a fist holding a coin. Then the fist disappears and the coin appears. At first, the image is not stable. After it becomes clear, concentrate your mind on punching a hole. When the intention of drilling a hole is concentrated to a certain degree, a transparent, crystal-like hexagonal rod will appear on the screen. It feels like it is very hard. As the intention of punching a hole persists and strengthens, a round tip suddenly pops out from one end of the hexagonal rod, and a hole is punched in the coin with a pulse. When punching a hole, the brain will feel a shock, as if the screen is about to be shattered. At this time, open your palm and there will be a hole in the coin."*

However, while all this was happening, people in our space could not see or sense anything. It seemed that everything was happening only in the space that only Chulin could access. Perhaps this is why many experiments with special abilities were denied. Because everyone only saw a result, and not the process, people would inevitably suspect that it was cheating. However, since it is a special ability, should we use a special way of thinking to look at it? If we jump out of the normal thinking framework, perhaps we can get in touch with a wider and more magical world.

So, did she cheat? Many scientists who participated in the experiments denied that she cheated because these experiments were conducted under their close supervision.

The hexagonal rod in Sun Chulin's experiment existed in another dimension, but it punched a hole in the coin in this dimension. Isn't it amazing?

Master Li Hongzhi revealed:

> *"Various masters have claimed that they can use their minds to do superhuman things like moving physical objects, opening the inner eye, or doing healings. They say that because they aren't aware of the powers they have. They only know that things happen for them at the thought of it. What's actually happening is that their powers are acting as the command of their thoughts, Which are in the brain; the man alone isn't making anything happen. So some of the things practitioners accomplish are due, in fact, to the powers the possess. Supernormal powers are innate abilities of the body, only they have declined as a world has developed and people's minds have become more complicated, with the "here and now" becoming increasingly valued and people relying more and more on the latest technology. To recover you innate powers, you must strive for authenticity as you practice, become pure as your once were, and restore your original nature-or what's the Daoist tradition refers to as "returning to one's original, true self." So what are now thought of as psychic or supernormal powers are in fact the innate abilities of man." (Zhuan Falun, p369-370, 1994, Li Hongzhi)*

Communicating with plants

Sun Chulin also published an article in the *Chinese Human Science* magazine on how to communicate with plants. It feels like reading a fairy tale. Let me share a few with you. *Chinese Human Science* was a monthly academic journal sponsored by the Chinese

Human Science Society in the 1990s. It was an authoritative journal in the field of special ability research at that time.

Chulin said the most important condition for the success of the experiment is to have spiritual communication with the plants. After communicating, you will find that plants have emotions and spirituality. They will share their joy, anger, sorrow, and happiness. Their voices are clear and they all have personalities.

Once, when experimenting to accelerate the germination of soybeans, a soybean complained, "It's too crowded! Too crowded! I can't stand it!" Sun Chulin didn't understand what he meant at the time. After the sprouts grew out, she found that the bottle was too small and there were too many beans in it. The bean sprouts grew into a mustache, which was funny.

Another time, she experimented with a peanut. The experiment was not successful. The next day, Sun Chulin communicated with the peanut, and the peanut began to speak, saying he was uncomfortable and in pain. Sun Chulin used her third eye to take a closer look and saw a thin wire through the peanut. It turned out that the professor who asked her to conduct the experiment had secretly marked the peanut in advance in case it was swapped.

Another time, a bean helped Sun Chulin point out the mistakes in the experiment. The professor gave her three red beans for the germination experiment. But Sun Chulin kept muttering: "Mung beans, mung beans, sprout quickly! Mung beans, mung beans, sprout quickly!!" Then she heard the bean respond: "Wrong! Wrong!" She didn't understand, what's wrong with me? Are you talking nonsense? The bean said

"I'm not talking nonsense, you are really wrong." Sun Chulin didn't insist on her own opinion. After a closer look, she found that she read it wrong.

Some scientists have discovered plants and matter are alive, can speak, and have thoughts. As *Zhuan Falun* says:

> *"Animals and plants count as lives, just as humans do; everything is alive in other dimensions. When your inner eye has reached the level of the Dharma eye, you may find that stones, walls, or any object might speak to you and greet you." (Zhuan Falun, p284-285, 1994, Li Hongzhi)*

Peanut resurrection experiment

Dr. Lee Si-Chen and a group of scientists witnessed the resurrection of the peanut. When peanuts are fried and die, the molecules such as protein, enzyme, or DNA of the cells dissociate and deform. However, the smaller particles that make up the molecules may not be destroyed. People with special abilities can use consciousness to control the process of molecular dissociation, which is equivalent to turning back time and restoring the damaged molecular structure to its original state. Peanuts are resurrected.

Sun Chulin agreed with Dr. Lee Si-Chen's statement and she used consciousness to control the peanuts to restore their vitality. The inside of the peanuts would produce a spiral rotation from the outside to the inside, which she called reverse rotation. When the consciousness is used to ripen the nuts or promote germination, some substance, which may be a kind of information or energy, will rotate from the center to the outside in a spiral, sometimes fast, sometimes slow, sometimes sparse, sometimes dense, sometimes loose, sometimes tight. The places it passes through will immediately produce obvious changes and the growth will accelerate as it rotates from the inside to the outside. She called it positive rotation.

Dr.Lee Si-Chen was shocked by the peanut experiments. If life can be reversed and time can flow backward how should we view death? In ancient China, the body must be kept for three days after death, which is said to give people a chance to be resurrected. Due to the development of online media, more and more cases of near-death experiences of people who were declared dead and then came back to life have become known to the public. Is this near-death experience related to this experiment? The Bible says that Jesus can resurrect the dead. Is this also related to the experiment they are doing?

Dr. Lee Si-Chen began to think about how "dead" a person must be considered truly dead. Of course, not all peanuts can be resurrected. Sun Chulin said that she had to communicate with them. For example, some peanuts would say that they didn't want to take root and sprout to produce a lot of small peanuts, and they were willing to be eaten by humans. In that case, Sun Chulin would not force them.

Dr. Lee Si-Chen pointed out, in his lecture at the Swiss Fa Conference held in Geneva on September 4th and 5th, 1998, that Particles are alive and have lives.

Removing the tumor with bare hands.

In the 1980s, the Qigong treatment method of grabbing was popular in China. It was said that if you grab the disease out it will be cured. Sun Chulin said she used the moving Qigong to grab a napkin and use her mind to grab a tumor out. After a while, the tumor with blood was wrapped in the napkin.

Like many special function experiments, people can't see the grabbing process, but the result is obvious. However, Chulin said this method is only done when she is in good condition. Grabbing consumes energy, there have been times when Sun Chulin fainted after treating others. She had to meditate and practice to recover, so she has only grabbed a few times in many years.

Master Li Hongzhi once explained that the phenomenon of Qigong masters being physically weak after treating patients is because:

> *"Energy like chi does not heal. And another thing is, a person will be connected with whoever he tries to heal, and it's possible that the sick energy in the recipient's body will come over to the healer's, such that he ends up with just as much of it. Though the root of the person's ailments won't be transferred, the healer might still become ill if he takes on too much of it." (Zhuan Falun, p89, 1994, Li Hongzhi,)*

CCP, which supported Sun Chulin in developing human special abilities, listed *Zhuan Falun* as a banned book. Because the book was banned, Ms.Sun Chulin did not have the opportunity to read the book, which put her health in danger when she used *chi* to treat patients.

Turn back time

What is most shocking about Sun Chulin is her ability to "resurrect from the dead." We have previously introduced that cooked peanuts can be revived and sprouted by holding them in her hands and squeezing them. She has not only done this experiment in Professor Shen Jinchuan's lab but also with Dr. Lee Si-Chen. The experiments are repeatable. According to her, she has done it thousands of times with a high success rate. However, this is not the ceiling of Sun Chulin's superpowers. In an experimental report circulated on the intranet (There is no Internet in China because the government controls people's freedom of speech.), there is even a record of Sun Chulin bringing boiled quail eggs back to life. In an interview in 2022, the host asked about this Incident and Chulin said," Yes, I have done it, it's very simple, just think about turning back time and returning them to their uncooked state, this is the case with peanuts and eggs."

Chulin said she even experimented to bring braised shrimps back to life. To challenge her, a seafood merchant picked some live shrimp from his shrimp farm and took them to a restaurant he designated to have the chef cook them. Sun Chulin claimed, "Living shrimps have souls, and after death, these shrimp spirits float in the air, as if they are reluctant to leave their bodies". Chulin explained that she communicated with the floating shrimp spirits, caught one, and let it go back, and a shrimp turned green. She caught another one and brought it back, and another shrimp began to jump around. At that time, the seafood merchant didn't quite believe it, so he tasted the shrimp that came back to life and validated that it tasted the same as the fresh shrimp sashimi he had eaten for decades using his shrimp.

Even more incredible is that she can apply these abilities to humans. Qigong was very popular and Sun Chulin said she had also done it. In a video that is said to be unavailable for broadcast in mainland China, Sun Chulin said that she once used the method of "turning back time" to make a Japanese child's tumor disappear. She said that she made the child go back to the time before the tumor developed, about 3 years old.

Then the question arises, children must grow up, right? They can't just stay at the age of three. Will the tumor grow back, and the disease recur? Sun Chulin said it is possible, but it may depend on the child's diet and lifestyle. This answer sounds very sincere. Indeed, many people who have been cured by qigong have relapsed, which is one of the main reasons why the qigong craze gradually subsided. However, many qigong masters will not admit it so clearly, but this "Superwoman" is very frank, many people choose to believe what she says. Master Li Hongzhi had the answer to the situation that puzzled Sun Chulin and explained:

> *"Then what explains all of the energy healing that's being done and taught out there? In most cases is that the people involved have gone down the wrong road. True Masters of energy practices like chi-gong, by contrast, develop compassion in the course of their spiritual progress and become sensitive to people's suffering. And so it is allowed, when that's the case, for master to help someone if he feels moved to do so. But what he does won't cure the person. He can only mitigate the ailment for the time being, or push it off to a later date in the future, or he might transfer it to someone else in the person's family. He can't thoroughly dissolve the karma for the person. You can't really do that for a regular person; It can only be done for people engage in spiritual practice. So there are principles at work in all of this.(Zhuan Falun, p76, 1994, Li Hongzhi)*

It is worth pointing out at the height of the Qigong craze in China in the 1980s and 1990s, according to statistics, there were more than 2,000 kinds of Qigong in China, but as of 2024, only Falun Gong is still widely spread around the world. Falun Dafa has experienced more than 30 years of persecution by the

CCP, which no other Qigong has experienced. Not only has it not been eliminated by the CCP, but it has also enabled people of all walks of life from different nationalities, beliefs, and cultural backgrounds in more than 150 countries to experience the great benefits that the universal law of truthfulness, compassion, and forbearance brings to people's physical, psychological, and spiritual levels. Other Qigong schools have not experienced the persecution of the CCP and have gradually faded away. This is true blue will never stain!

Sun Chulin has 60 superpowers. Due to the length of this book, I only introduced a few of the most incredible ones here.

IV. Why Can't You See Divine Beings?

We have to go back to the finger reading experiment conducted by Dr. Lee Si-Chen.

Dr.Lee Si-Chen: "This is a historical record. This shot is a historical record. It is the first time that human beings discovered the image of Buddha!"

At 2:30 p.m. on August 26, 1999, a group of more than a dozen professors from the Department of Physics and the Department of Psychology from National Taiwan University, National Tsing Hua University, and Academia Sinica marched into Conference Room 142 of the Second Building of Electrical Engineering at National Taiwan University.

Tang Dalun (a psychologist): "So I originally participated with the mentality of wanting to expose the scam."

The subject was a 17-year-old Taiwanese-Japanese girl, Mai Takahashi or Gao Qiaowu by Pinyin.

Dr.Lee Si-Chen: "These are instruments attached to her hands, detectors, which are electrodes, one on the left hand and one on the right hand. These two electrodes are sent to what is called a strip chart recorder. The paper on it is constantly moving, recording the voltage of her two hands."

Staff: "So when the voltage changes, does it mean that her third eye is open?"

Dr.Lee Si-Chen: "Her third eye opened two seconds after the voltage changed, so when we saw the voltage, we knew that her third eye was about to open. After one second, it opened and she could see part of it, so at this time she would start to talk and we needed to record what she saw."

The method of the experiment is simple. Anyone at the scene can write a random word on a piece of paper, fold it, and put it in a dark bag. The subject can read the answer by touching it with her fingers.

But the surprise is yet to come because a physics professor wrote a special word and put it in.

In the experiment:

Dr. Lee Si-Chen: «There is a what?"

Takahashi: "There is a person on it!"

Dr. Lee Si-Chen: "Did you see it? What does this mean?"

Takahashi: "This word is more complicated."

Dr. Lee Si-Chen: "Can you describe what you saw to me?"

Takahashi: "I saw the shape of a person. There was a shadow on top." (of the word she was reading/ touching with her finger)

Takahashi: "There is a light next to him."

Dr. Lee Si-Chen: "What does he look like?"

Takahashi: "I can't tell?"

Dr. Lee: "Is he smiling at you?"

Takahashi: (nod)

Dr. Lee: When we saw the bright light and the bright people, we felt very strange. Then we opened it and saw that it was the word "Buddha". This was the first time we discovered that the word "Buddha" could produce images. Everyone was shocked, including me.

In the experiment: "What she saw was not words, but the image of Buddha." "Where is it?" "Because Buddha looks like your master, very pious and kind, smiling at you."

Before this, Dr. Lee Si-Chen had been doing experiments with Takahashi for six years. He had never seen such a result. An unintentional attempt unexpectedly opened a door to an unknown world. Even Dr. Si-Chen could not suppress his excitement.

Dr.Lee Si-Chen:

> *"I tried all kinds of methods, but I never thought of trying these sacred words of religion. I never thought of this idea at all. So, when the first word "Buddha" came out, everyone saw it, including the professors from the Physics Society. I secretly said*

in my heart, it's good, people didn't believe it in the first place. If I did this at that time, it would be even more sinful. Now it's not like that. It was the Physics Society who gave it (the word "Buddha"). Seeing it is their credit, not mine."

Dr.Lee Si-Chen:

"Yes, a completely new research direction, and it also began to make us understand what the truth of the universe is. In the experiment, "this world exists, it should be called the spirit world. I don't call it the spirit world, but the highly intelligent information field. This thing must exist. From the experimental results we have seen, because the results just came out, we have to burn incense and start worshiping Buddha (laughs)"

The shock brought by this discovery changed all of Dr. Lee subsequent experimental directions and plans. Is the bright human figure that Takahashi saw from her hallucination or external stimulation? Did Takahashi see the word Buddha?

Dr.Lee Si-Chen:

"So many people had doubts about this and thought that she originally had respected this kind of thing very much, so when she saw this word, she started to have many hallucinations like dreaming. This was a very good question, so we must understand whether it was her hallucination or there was really another world. We tried all different words such as Jesus and the Saint Mary in English. All Tibetan characters of Guru Padmasambhava, and the two-armed golden-robed Vajra, all produce images.

As long as the characters were correct, images would be produced. In order to verify this, I had deliberately cut off a part of the character which caused her not seeing the image but only the cut part of the words.

So, from this point we can understand very well that the character is a concept of a URL. It is a network world, so these characters are all URLs of websites in the information field. We understand that it is very important, so we can go to various religions to see their websites.

The scientific basis behind Dr. Lee Si-Chen 's theory was clearly shown in the brain MRI of experimental subjects. Dr. Lee said that whenever the words hidden in the bag are religious, this gratifying brain activity would become more frequent. The changes in the posterior part of the callosum were generally 3% to 5%. However, Takahashi's was 40 times larger. Dr. Lee believes that the mystery of the spirit medium might be explained by his discovery. As long as the medium has excellent extrasensory perception, he or she can build a bridge between the secular world and the divine world and thus obtain the oracle.

"Unexpected" Happened During the Experiment

The psychologist Tang Dalun wrote the word "ghost" for Takahashi to touch. After she touched it, she felt goosebumps all over her body; her third eye, which could have opened and lit up, was covered by a black shadow. Dr. Lee saw the word and wondered who had prepared it. Takahashi was shocked, she felt something was wrong the whole day and could not continue with the experiment. She thought the world was scary.

Reporter: "If you input Satan, the devil?"

Dr. Lee, "In our laboratory, it is forbidden to experiment with these things. It is not allowed to touch them. This is an ethical issue. Therefore, our laboratory strictly prohibits touching these things. The characters and the words for testing must be positive because you don't know what will happen. You can't do such things that harm children. So only scary words with positive energy and positive things in religion are allowed for testing."

On July 13, 2024, former U.S. President and presumptive candidate for the 2024 Republican presidential election, Donald Trump, escaped an assassination attempt while speaking at a campaign rally near Butler, Pennsylvania. Anti-Trump atheists believe that it was just Trump's good luck and a coincidence.

So why could psychics like British psychic Craig Hamilton-Parker, Texas Vedic astrologer Joni Patry, Oklahoma pastor Brandon Biggs, and other psychics predict in advance that Trump would survive an assassination attempt?

Pastor Biggs was even able to provide details and accurately predicted the assassination attempt four months before the event. "I saw Trump stand up, and then I saw someone trying to murder him," Biggs told Pastor Steve Cioccolanti. "This bullet went by his ear and was so close to his head that it broke his eardrum," he said, mentioning that he saw this scene a few weeks before the recording. The YouTuber stated he witnessed Trump fall to his knees and begin to worship God.

I believe that those atheists cannot explain why so many people could foresee the failed assassination attempt on Trump. Moreover, there has been more than one assassination attempt on Trump, but none of them succeeded. Who has been protecting Trump? I think you should know it after reading this chapter, right?

Dr. Lee Si-Chen, Qian Xue-sen, and Sun Chulin have been dedicated to the scientific and official experiments that can be repeatedly verified. The conclusion they all came to is that divine beings live in the same space as us, just in different dimensions. We cannot see them with our naked eyes, like we cannot see ultraviolet rays and sound waves.

To be precise, *Zhuan Falun* has pointed out that we humans live in the third dimension, which is the lowest dimension that can sustain human life, and gods can survive in any dimension higher than us, of course including the third dimension. That is why *Zhuan Falun* says there are gods, Buddhas, and Taoists throughout the universe, including the dimension we live in. This is consistent with Dr. Lee Si-Chen's research results indicating there are gods, Buddhas, and Taoists everywhere around us.

William Mueller, the most famous soul photographer in American history, once successfully photographed the First Lady Mary Todd Lincoln and the soul of President Lincoln. In Muller's photo, President Lincoln's white figure affectionately stood behind his wife, looking at her with infinite love, with his hands on her shoulders, which was very touching. Experts determined that it was difficult to synthesize this photo using the double exposure technique because the president's body was behind the First Lady, but his hands were in front of her. If the double exposure technique was used, it had to be synthesized twice. Without a computer and unable to retouch the photo, it is difficult to achieve the natural look in the photo, with no seams visible. Is it the soul of President Lincoln in the photo? I am not qualified to explain this knowledge to you because no human being understands this knowledge but if you read *Zhuan Falun* from outer space, you will learn a lot about the spiritual world and the truth of the universe.

Why You Can't See Divine Beings?

I would like to quote Master Li Hongzhi's explanation below:

> *Let me illustrate it. Spiritual traditions have taught that the phenomena of the human world are illusory, or not real. But how so, one might wonder? Who would say that a solid object right in front of you is not real? The way that objects appear is different from how they actually are. But our eyes have the ability to stabilize the appearance of objects in this dimension so that they look as they do. But that isn't in fact their true state, not even in this dimension. For example, what would a person look like under magnification? The whole body would seem loosely comprised, with molecules moving all about that are small and round, like grains of sand. Electrons would be orbiting around atomic nuclei and in motion, with the entire body pulsing and moving. And the surface of the body would be irregular, not smooth. The same holds true for any object in the universe—even iron, steel, and stone. The molecules that make up all things are moving, only you can't see the full picture. Nothing is in fact static. Even this table before me is moving, only our eyes can't perceive the reality of it. Our eyes can mislead us."(Zhuan Falun, p53, Li Hongzhi)*
>
> *Contemporary science is capable of seeing the composition of the largest molecular particles. A feature of steel is that it's stronger than iron, because the elements and structure of its particles are different. What is copper like? What are silver and gold like? Science can grasp these things. So then it utilizes the materials known at present, and only develops with that as a basis. But let me tell you that that's the greatest obstacle to*

contemporary science in its effort to know other dimensions. In other words, it hasn't taken the path of understanding matter in microscopic worlds or in worlds that are even more immense. In fact, matter is made up of different small molecular particles, and those molecular particles are made up of atomic particles. Atomic particles are then made up of nucleic particles. And nucleic particles are made up of more microscopic particles. The more microscopic particles are made up of even more microscopic particles. The even more microscopic particles are in turn made up of even more and more microscopic particles. To humans it appears infinite and endless. It keeps progressing further down like this. How microscopic does it get? I can tell you that the microscopic state that particles reach is astounding!"(Teachings at the Conference in Europe, Li Hongzhi, May 30–31, 1998 ~ Frankfurt, Germany)

Master Li Hongzhi gave an example at the 1998 European Fa Conference. He said that the largest particles humans can see are not molecules, but planets. However, planets are not the largest particles in the universe. They are just particles between the macroscopic and microscopic. Just think about how many huge spherical particles there are.

In Frankfurt, he also told everyone that there are lives on all planets, and many of them are human-like lives, but they are not in this dimension, because different particles make up different dimensions. If you want to enter that dimension, you cannot use the current human science to understand it, because this science is too superficial. If you enter that world, you must conform to the state, form, and thinking of that world, and then you can understand it. After you enter you will find the particles are smaller than the particles of the human level, but its space is vast. We humans live between two types of particles: molecules and planets.

So, according to Master Li Hongzhi, we humans live between the two particles of planets and molecules. Any particle larger than a planet or smaller than a molecule cannot be seen by our naked eye unless our third eye is open. Therefore, if we put a lot of bacteria or viruses that are even smaller than bacteria into a human shape or a larger object shape, our naked eyes cannot see the human shape or other shapes formed by the bacteria or viruses. This also explains that no matter how many and how densely there are bacteria and viruses in our space, our naked eyes still cannot see them.

Before the invention of the microscope, human eyes could not see bacteria. Now we know that viruses are smaller than bacteria, and there are particles smaller than viruses. The world of divine beings is made of small particles, so we cannot see life in other dimensions, including the world of gods, Buddhas, and Taoists.

We cannot see these bacteria and viruses with our eyes, but they exist, right? Now, Dr.Lee Si-Chen has proven through rigorous scientific experiments that God's world exists, even though we cannot see it. Thus, if you still firmly believe that seeing is believing, then I think it is not a question of whether God exists, but a problem between you and God.

V. Is There Only One God in the Universe?

Master Li Hongzhi often points out that there are countless gods, Taoists, and Buddhas in the universe. This seems to have been proven by Dr. Lee Si-Chen through scientific means. I think this is a shock not only to atheists but also to monotheists.

I used to wonder why Elohim told the Jews that he was the only God in the universe and the only way for the Jews. Jesus told Christians that he was the only way for Christians. Buddha Shakyamuni told his disciples that he was the only way, many Buddhas in Buddhism have said this to their disciples. Then, who is the Lord of the universe who created all things? Whose way is the genuine embodiment of the way that can save people?

An enlightened disciple of Master Li Hongzhi once gave me some inspiration. I think we should look at this issue in this way.

Suppose as a Christian, you knew that Jesus bought a house on the West Coast of the United States, and you decided to go to Jesus' home. The moment you entered the address of Jesus' home into your navigation and decided to take the *way* to his home was when you decided that the *way* to Jesus› home was the only *way* to save you and that Jesus was your only Lord.

If you were in Texas, you could not take the road to the east coast of the United States at this time, because that road might take you to the home of Buddha. But Buddha did not know you and he would not accept you. As a result, you would not be saved.

At the same time, you could not choose any path other than the way that led you to Jesus' house, because this universe was mixed with gods and demons, and any path you took may lead you to the home of Satan or other demons. The result was that not only would you not be saved, but you would also go to hell.

This universe is dangerous, so whether it is Elohim, Jesus, or Buddha, they have repeatedly emphasized to their believers that they are the only way for believers to be saved and the only God in this universe who can save them.

So why do Elohim and Jesus say that they are the kings of all kings in this universe?

Master Li Hongzhi once answered this question. He said that there are more than 100 such heavens in the Milky Way galaxy where we live, and each heaven is presided over by a great enlightened being. So, I think there should be different kings in each heaven of the great enlightened being.

Our knowledge of Jesus and God comes from the Bible, but the truth is that the Bible is a book of prophecy, history, and the origin of humanity, and it is also a collection of disjointed and incoherent words that people recall as being said by God. The people who wrote the Bible at that time came from different cultural backgrounds and different historical periods. Therefore, some translation errors, interpretation errors, and memory errors are inevitable. The scenes, places, and people in which God spoke at that time were all different from today. Most of the people who listened to Jesus' sermons at that time were shepherds, farmers, and those people who had not received modern higher education, and many of them were even illiterates. When Jesus preached to them, the language he chose was extremely limited too.

The differences between the biblical stories and historical reality caused by human factors are intentional by the divine world. For instance, Master Li Hongzhi explained because the divine world did not allow Jesus to write the Bible, it had to be written by humans because humans make mistakes and cause the biblical verses to appear confusing and contradictory. In this way, God can see who still firmly believes when people are confused and contradictory. This can be confirmed from the story of Job in the Old Testament. Job was a rich man with a happy family, and he was very devout to God. One day Satan was trying to sow discord between Job and God. Satan told God that Job was devout to God because God gave Job everything, so God began to test Job's faith in Him. Job lost his wealth, his family, and even his personal health. Job was confused, he didn't understand why God treated him like this when he's so faithful to God. But Job did not complain to God, instead his faith in God was as solid as a rock. In the end, Job passed the test, and God gave everything back.

The Buddhist scriptures were also written by humans. Why didn't the Buddha write the Buddhist scriptures? Because he was not allowed to,

This is why the Bible has caused a lot of confusion to mankind today, leading to many pastors interpreting the scriptures randomly according to their understanding.

After I finished reading *Zhuan Falun*, my confusion about the Bible and religion was finally resolved.

So, what did Jesus tell people?
My understanding is that Jesus only told people two things.

First, Jesus told us the truth about life.

Our souls are who we really are, and they are not meant to be on earth. One of the functions of our flesh is to punish and restrict us. Why does God use the flesh to punish and restrict us? It is because we were all "criminals." This is why Jesus always said that everyone is sinful. The earth is a prison and our bodies are used to confine us.

"In other dimensions people do not have bodies like those here, and they can float in the air and enlarge or shrink. In this dimension, by contrast, people are saddled with the physical bodies that we now have. With this kind of body, cold or heat will be unpleasant, and it's difficult when you are tired or hungry. Pain is inevitable. There are the miseries that come with illness, and you are subject to the cycle of being born, aging, becoming infirm, and then dying. All of which is meant to pay off your karma, through suffering. You are being given one more chance to go back, and that is why you are in this beguiling place. After dropping here, you are furnished with eyes as you now have, which prevent you from seeing other dimensions or seeing the material world for what it is. Those who make it back, however, will find that the worst of their suffering here was in fact the most valuable. Practicing by relying on faith while in this confusing place is extremely trying, yet it is also what makes a swift return to heaven possible. But if instead you become still worse here, then your soul will be extinguished. So the point of human existence, as higher beings see it, is not to be human but to recover our original purity and go back to our heavenly home. The average person doesn't realize this, however, for this secular world is all he knows, and what tends to occupy his mind is figuring out how to get ahead and live the good life. But the better off he is, the more selfish and greedy he may become, in which case he strays only further from the qualities of the universe and heads for destruction." (Zhuan Falun, p113, Li Hongzhi)

Moreover, our body has many needs. To meet these needs, we must work non-stop. Isn't work hard? Especially when your job is something you don't like to do, it exacerbates our pain.

Second, Jesus taught people a scientific method on how to achieve the purpose of life.

Because only by getting rid of this flesh can we avoid birth, aging, sickness, and death. Therefore, the scientific method that Jesus taught people to achieve the purpose of life is to tell people how to get rid of this flesh that tortures and binds people's souls to achieve eternal life, that is, to return to the state before we came to the earth.

According to Master Li Hongzhi, this scientific method is called cultivation. Every great enlightened being has a unique scientific method that enables his followers to achieve perfection in cultivation. For example, in Christianity, Jesus told people to do things according to what he taught them, pray, repent, and not make the same mistakes after repentance to improve their character and meet the requirements to enter a higher dimension.

Falun Dafa is also a cultivation method. That is why Master Li Hongzhi repeatedly emphasized that Falun Gong is not a religion but a higher science that cannot be understood by our modern science.

Einstein said: "Science without religion is lame, religion without science is blind."

What Einstein said about religion is not accurate, because Master Li Hongzhi once pointed out that God did not teach religion to people. Religion is created by people themselves. When Jesus was here, Jesus did not say what he taught to mankind was called Christianity. There was no Bible or church at that time. When Sakyamuni Buddha was alive, Buddha did not tell people that the knowledge he taught was called Buddhism. There were no Buddhist scriptures or temples at that time. It was humans who created various religions themselves.

As for what Einstein meant by science, that is, what exactly is modern science and what role does the existing science play in the pursuit of immortality? We will analyze it in the next chapter.

CHAPTER 2:

Modern Science and the Truth about Immortality

In the first chapter, we learned that Dr. Lee Si-chen could verify the existence of gods and different realms by scientific means. Ms. Sun Chulin could communicate with divine beings in the space where gods live and learn supernatural abilities. So, if humans want to pursue immortality, should we give up the idea of keeping their bodies forever young? Should we do the exact opposite and get rid of our bodies? Should we let ourselves live with those gods, Taoists, and Buddhas in a dimension not bound by the physical body, time, and space?

Is it true that humans want to achieve immortality by keeping their bodies young forever or is it a result of being brainwashed by human science? To answer this question, we must first understand what science is. What is the true meaning of immortality? And why do you have a body?

I. What Is Science?

Sadhguru, an Indian spiritual leader and yoga instructor, can solidify mercury with his bare hands, while modern science believes that mercury cannot be solidified. But Sadhguru's superpowers are nothing compared to Sun's more than 60 special abilities. But even Ms. Sun's many superpowers are insignificant compared to many Falun Gong practitioners. I am not here to belittle or praise certain people or groups with superpowers that cannot be explained by existing science. I just want to tell you that human science is limited and is not the real science of the universe. This kind of human science cannot even answer the most basic and important questions, such as where we came from, why we are on Earth, and where we will go in the future, let alone understand the structure of the universe.

The existing science of mankind is not suitable for unlimited development and application on this earth. You all know that under the promotion of our existing science, the water, soil, and air on which we depend for survival have been seriously polluted. The natural resources have been overly depleted causing irreversible environmental damage. Not to mention that in pursuit of the material enjoyment provided by science, human morality has fallen to the lowest point in history. I am not against human science, I just want to tell you that material life should slow down or even stop when it develops to a certain level. Our spiritual civilization has been left far behind.

Since human science is limited, human science is constantly exploring and understanding the universe. In the past, the mainstream scientific community once believed that the earth was flat. Today, is this idea ridiculous? In fact, there is no need to laugh at the scientific consciousness of the past, because the current mainstream scientific community is not much better than before. The best example is the double-slit interference experiment in quantum mechanics and its variants, which subverted the cognition of the physics community.

Why does a single photon become two photons after passing through the interference screen with double slits? Have you ever seen a circus lion jump through two fire hoops side by side at the same time? What is even more incredible is that depending on whether an observer is added to the experiment, the photon can also choose what form to show itself to you. Isn't this equivalent to when you supervise your children, they work on their homework? When you do not they are all on TikTok. Are photons alive and conscious? Human scientists only use wave-particle duality and the observer effect to explain phenomena they cannot understand.

Japanese scientist Dr. Masaru Emoto conducted thousands of water crystallization experiments and published a book called "Water Knows the Answers". The experimental results prove that water has emotions just like people. Water that is praised will produce beautiful crystals; water that is scolded will produce ugly crystals. Water can even understand human language, read the meaning of words, and appreciate music. According to the experimental results, water can also understand different languages. No matter which country's language is used, if the word "gratitude" is posted on the glass bottle, the water crystals will show beautiful patterns. Can water molecules also have life?

In fact, Master Li pointed out in the Fa lecture at the Swiss Fa Conference held in Geneva on September 4-5, 1998, "*Particles are alive and have life. So think about it, no matter how big the universe is, it is alive. But within the scope of celestial bodies, there are countless particles of different sizes, and they are also living beings.*"

Why did Einstein start studying theology in his later years?

In Master Li's words, Einstein was smarter than other scientists. In 1998, Master Li Hongzhi said at the European Fa Conference:

> *Why did Einstein turn to religion in his later years? He was so accomplished in science, but he found that religion was right. People don't understand him because they are not as smart as Einstein. Those stubborn people don't understand, and those who are not as wise as Einstein don't understand.*

Albert Einstein also said, "The more I study science, the more I am amazed by the complexity of the universe and the more I believe in the existence of a creator." In other words, if a scientist doesn't believe in God, he or she doesn't study science enough!

Theology is the end of science. Does this sentence sound reasonable now?

Today's science is so ignorant that it believes that what cannot be seen and understood does not exist and is not science.

If people cannot see viruses and bacteria, does that mean the plague does not exist? I think everyone still remembers the global plague caused by the COVID-19 coronavirus that leaked out of the CCP laboratory at the end of 2019. 500 years ago, when there were no microscopes, if someone told the public that the plague was caused by viruses and bacteria, he might be burned to death as a wizard. Now when you say these things to the public, no one will regard you as a wizard, because this is common sense. Therefore, the scientific framework of mankind has been expanding in continuous learning and exploration. It is very necessary to maintain an open attitude towards areas that science has not yet explored.

The tests that Master Li Hongzhi underwent showed that the gamma rays and thermal neutrons he released were 80 to 170 times higher than the normal amount released by matter. This was only the amount that the equipment could measure because the indicator had reached the limit. Both gamma rays and thermal neutrons are invisible to the naked eye.

I would like to ask how to explain these people with superpowers with science. In the words of Master Li Hongzhi, people who believe in modern human science are a group of scientific superstitious people. Wanting to keep your body in a state of immortality is a kind of superstition. *Zhuan Falun* says a person's life and how long each person can live is determined. Einstein also said, "Everything is predetermined, including the beginning and the end."

True eternal life is the method that Jesus and Buddha gave us to free ourselves from the flesh that binds our souls so we can live without the constraints of time, space, and physiological needs.

In fact, the culprit that caused you to misunderstand the definition of eternal life is Western modern science. Among the Falun Dafa lecture documents, in an article called *The Confines of Modern Science and the Breadth and Profundity of Buddha Fa*, Master Li Hongzhi argued:

> *Someone has defined science, and only those that develop according to it are considered science. Before you go beyond its scope, you think it is science. When you go beyond its definition, you find that it is limiting human development. It does not recognize anything that cannot be touched*

or seen, and its limitations are very large. The Buddha, Taoism, and Gods we talk about are all beings in other dimensions that people cannot touch or see. If they are explored using current scientific methods, then are they scientific? They are also science! However, the West has already defined science, and it attributes everything that modern science cannot explain to theology and religion. It dares not recognize it."

Moreover, in another published article-*Further Comments on "Superstition"* Master Li pointed out:

Whether something is science or superstition isn't to be decided by politicians. It should instead be appraised by scientists. Yet the so-called "scientists" being used for political purposes are in fact political figures. Such people can't draw fair and scientific conclusions from a truly scientific and neutral standpoint. So, ultimately they can't be considered scientists. At most they are tools in the hands of politicians, used to attack people.

The two articles written by Master Li Hongzhi explain very well why Dr. Lee Si-chen discovered and verified the existence of gods, Buddhas, and Taoists with rigorous scientific methods, but he was not awarded the Nobel Prize.

II. Western Science Lacks Prehistorical Knowledge

Master Li Hongzhi mentioned:

There are many artifacts around the world that far predate the earliest era of civilization now known. Many of the remains display a sophisticated level of craftsmanship and artistry. The arts of the past had a high aesthetic value and almost give the sense that modern man was just following in their footsteps. And yet some of these remains date back hundreds of thousands or millions of years, with some even exceeding 100 million. It would seem to radically challenge current views of history, wouldn't it? But there is nothing inherently radical about it. People always try to push the boundaries of what is humanly possible, and keep arriving at new insights about themselves and their past. So people's initial findings and views about things aren't always right.

This leads to the topic of prehistoric civilizations, or prehistoric cultures, which you might be familiar with. In the world today we have Asia, Africa, North America, South America, Antarctica, Europe, and Oceania, which are all considered in geology to be continental land masses. These settled in their current configuration tens of millions of years ago. So some land masses emerged from the ocean floor, while others, now no longer with us, became submerged under the ocean's waters. The current land masses have been in place for tens of millions of years. Yet discoveries have yielded ancient megastructures at the bottoms of several oceans. And these structures are covered in intricate carvings bearing no resemblance to the works of any civilization today. Yet they must have been built before having sunk to the bottom of the sea—many millions of years ago. But then who would have built them, if man, supposedly, wasn't even a monkey yet? Such sophisticated things wouldn't have been possible. Another enigma surrounds the discovery by archaeologists of a certain trilobite fossil. This is a creature that dates back some 260 to 600 million years, and has long been extinct. Yet an American scientist discovered a trilobite fossil with a human footprint on it—the footprint of someone wearing shoes. And the print was unmistakable. For many, it's an embarrassing challenge to established views about the past. Darwin's evolutionary model can't explain the existence of human beings that long ago.

Another telling case is a rock in the museum of a national university in Peru that has a human figure carved on it. An analysis determined that the person was carved on it thirty thousand years ago. But the person wears clothes and a hat, has shoes on, and is holding a telescope and observing the stars. And while it might be hard for us to imagine people weaving clothes like those back then, it's even harder to fathom the presence of a telescope and the possibility that they knew something of astronomy. People have always thought that it was a European, Galileo, who invented the telescope, some three hundred years ago. So we are left to wonder how a telescope came about thirty thousand years ago. And there are many unsolved mysteries like this. Another is the cave murals that have been found in various countries including France and South Africa, as well as in the Alps. These were painted skillfully, in a very lifelike and vivid way, with the use of mineral pigments. But what's striking is that the figures depicted in them are dressed like people from modern times, with European-style clothes, like breeches. And some are holding what look to be tobacco pipes, while others have canes or wear hats. Art of that quality is hardly what you'd expect from our supposedly primitive ancestors.

Another example goes further back. Uranium ore is present in the African country of Gabon, but the country is not very developed and hasn't the means to refine it, and so they have exported the ore to more developed nations. When a plant in France imported some of it in 1972, they were surprised to find, after chemical analysis, that the uranium had already been processed and depleted. They thought it was rather strange, and sent over scientific and technical personnel to investigate onsite. And eventually scientists from many other countries went as well. Ultimately, what they determined was that what was thought to be the uranium mine was in fact a well-designed, large-scale nuclear reactor—one built better than could be made today. And yet the reactor was built 2 billion years ago and had operated for around five hundred thousand years. That's an extremely long time ago, and makes no sense within an evolutionary framework. And there are many anomalies like this, in fact. There have been more than enough scientific discoveries to rewrite the books. It's hard for people to be receptive to new ideas or findings if their views on things are set and have formed into a mode of operating and thinking. So they will shy away from the truth, even when it comes to light, or instinctively reject it. Owing to longstanding biases, hardly anyone in the sciences is systematically reviewing or trying to make sense of these anomalous findings. And so the theoretical frameworks aren't keeping pace. New findings continue to be written off as "pseudoscience," even if they are valid, though not well-publicized.

A number of scientists in other countries who are a bit more courageous have openly voiced their opinion that these finds do bespeak of earlier cultures prior to written history or civilization as we know it. So this would mean that they did indeed exist, and it's believed that there was more than one—judging from the artifacts that have been unearthed so far. It is now believed that on multiple occasions something catastrophic befell civilization and left the population decimated. The few people that survived would have lived a most basic existence at first, until, with time, the population had grown again and civilization developed anew. And this cycle would have repeated itself many times, with phases of destruction and renewal. All changes throughout creation follow patterns, much like how in physics the motion of matter follows patterns." (Zhuan Falun, p18-21, Li Hongzhi, 1994)

Master Li Hongzhi mentioned:

The current Western science originated from European science, and European science did not inherit any knowledge obtained from the previous human civilization, but was rebuilt. "Humanity does not know its historical origins, and the same is true for white people. In the last great flood, the last time human civilization was destroyed by a great flood. All the mountains below 2,000 meters above sea level on the earth were flooded, and only people living above 2,000 meters survived. The story of Noah's Ark is true. Western culture was completely destroyed in this great flood. Eastern culture was also destroyed, but people living in the mountains around the Himalayas and Kunlun Mountains, like rural people, survived, and the Chinese living in Kunlun Mountains survived. Because at that time, Eastern culture was very developed, it inherited the tai-chi symbol, the mystical he-tu and luo-shu diagrams, the Book of Changes, the Eight Trigrams, etc. People say that it was created by someone afterward, but that was not the case. They are all from prehistoric Chinese culture. Actually, China has been left with more prehistoric knowledge than these in the course of history. In ancient times, there were more prehistoric things left, but the knowledge from their ancestors became less and less as they were passed down. So, China really is a nation with

> *deeper connotations and a longer history compared to the West. After the white people's culture was completely submerged by the great flood, there was nothing left. At that time, there was a continent on the edge of the European continent, which also sank. The sunken European continent was the West's most developed place. So, the white people redeveloped from a state of having nothing and no civilization, which is their current science. " (Lecture at the New York Fa Conference, March 23, 1997, Li Hongzhi)*

I think this is why some ancient civilizations in the world, such as Jewish culture, Hindu culture, and traditional Chinese culture, know that a total solar eclipse is a bad sign because these ancient cultures have inherited some knowledge left by prehistoric civilizations. Many books in traditional Chinese culture record prehistoric knowledge, such as the *tai-chi* symbol, the mystical *he-tu* and *luo-shu* diagrams, the *Book of Changes*, the Eight Trigrams, etc., which are all left over from prehistoric times, according to Master Li Hongzhi.

I think this is why in the West, whenever a total solar eclipse occurs, people think it is an astronomical wonder and rush to see it. Astronomy enthusiasts will even celebrate the phenomenon as a festival. Little do they know that this is God's warning to the world, indicating that a major disaster is about to occur. On April 8, 2024, a total solar eclipse crossed North America, passing through Mexico, the United States, and Canada. Although the partial solar eclipse can be observed across the United States, the total solar eclipse only passed through a few states, including Arkansas, Indiana, Ohio, New York, and Texas. And more than 600 million people in the countries or regions where the moon's shadow passed saw the solar eclipse.

In this world, all ancient civilizations know that strange phenomena from the sky will bring disaster.

Background of Nineveh's Repentance

Everyone is probably familiar with the story of Jonah in the Bible. At that time, Assyria was a very powerful nation. The key to their rise was a highly efficient and brutal army full of bloodshed, which was used to humiliate their enemies. The Assyrians worshiped many pagan gods. Nineveh was the greatest city at that time with a large population. According to historical documents, Nineveh had a population of about 1.5 million and was a veritable Sodom. Have you ever wondered why Jonah's message of repentance to the Ninevites had such a magical effect? A set of cuneiform tablets discovered in the 19th century provides a clue. During Jonah's ministry, Nineveh experienced two solar eclipses. The first Assyrian eclipse occurred on June 24, 791 BC, obscuring 74% of the sun, and the second occurred on June 15, 763 BC, covering 98.7% of the sun, almost a total eclipse.

According to Assyrian documents, an eclipse meant that the king would be deposed and killed, a worthless man would seize the throne, rain from heaven would flood the earth, and the city walls would be destroyed. The Assyrians cried out, "Nineveh will be overthrown." Adad-nirari III became a scourge in the Middle East during his relatively short life, conquering a long list of peoples: Syrians, Phoenicians, Philistines, Israelites, Shinheites, Edomites, Persians, Nedi, Babylonians, as far as the Caspian Sea. The first of two eclipses occurred at the end of his reign. At that time, the Assyrian Empire entered a period of decline, experiencing a plague in 765 BC, a solar eclipse in 763 BC, a peasant uprising from 763 BC to 759 BC, and a second plague in 759 BC.

As Tom Hobson, chair of the Department of Biblical Studies at Morthland College, observes, the Assyrian state was weak and in disarray for 10 years around 760 B.C. They had a great earthquake, which was seen as a sign of God's anger. From 765 to 758 B.C., there was famine, Assyria continued to lose territory to its enemies, and there was unrest in the country. This was the perfect time for a prophet to come in and warn them. According to the Bible, Jonah walked into the city on the first day, and he shouted, "In 40 days, Nineveh will be overthrown." The king of Nineveh believed in God and declared a fast, and everyone, from the greatest to the least, was clothed in sackcloth. This was God's warning, so he stepped down from his throne, covered himself with sackcloth, and sat. He also sent word throughout Nineveh that the king and his officials had decreed that no one should eat any livestock or cattle or drink any water. Both humans and animals should be covered with sackcloth and cry out to God. Individuals turned from their evil ways and abandoned the violence in their hands, and the nation repented, saving Nineveh and the kingdom from its doom.

While it is generally believed that the 2024 total solar eclipse passed through seven cities named Nineveh: Pastor Steve Cioccolanti told Brandon Biggs on his YouTube channel "Steve Cioccolanti & Discover Ministries" that he counted the total lunar eclipse and found it passed through 11 cities named Nineveh.

Nineveh, Leon County, Texas; Nineveh, Adair County,, Missouri; Nineveh, Johnson County, Indiana; Nineveh, Darke County, Ohio; Nineveh, Greene County, Pennsylvania; Nineveh, Clarion County, Pennsylvania; Nineveh, Warren County, Virginia; Nineveh, Brooke County, New York; Nineveh, Nova Scotia, Canada; Nineveh, Victoria County, Canada; and Nineveh, Anderson County, Kentucky.

Is God giving the United States a final warning, just like He warned Nineveh? Nineveh experienced a plague after an eclipse, and then a second eclipse. Does this sound a bit like America today? After the second eclipse, Nineveh experienced civil war, a second plague, earthquakes, and famine.

On April 8, 2024, North America saw a total solar eclipse, which affected many cities in the United States. On the 5th, a 4.8-magnitude earthquake occurred on the east coast of the United States, with the epicenter about 60 kilometers from New York City. This is the largest earthquake in the New York area in 140 years. As I wrote this book in October 2024, natural disasters are becoming more and more frequent, and Americans hear bad news every day! However, there will always be some scientists who stand up and tell you that the total solar eclipse has nothing to do with earthquakes.

The Chinese believe that solar eclipses are a bad omen.

Since ancient times, people have believed that total solar eclipses are the wrath of God or a sign of the end of a dynasty and may indicate an impending catastrophe. There are records of major solar eclipses in ancient China. The ancients emphasized the unity of man and nature and observed the celestial phenomena to understand the earth. Ancient Chinese believed that solar eclipses were a bad omen, and ordinary people would panic when they saw a total solar eclipse. According to historical records, in the era when total solar eclipses occurred, the probability of unrest and war was 95%, and the probability of various natural disasters was also more than 95%. In modern Chinese history, there are four periods when total and annular solar eclipses occurred in large numbers, all of which were accompanied by major disasters:

* 1848-1857: three solar eclipses, the Taiping Rebellion, the British and French invasions, major floods, major earthquakes, and major locust plagues
* 1869-1875: four solar eclipses, major floods, major droughts, major plagues, and major earthquakes
* 1936-1943: three solar eclipses, the Anti-Japanese War, major locust plagues, and major famines
* 1965-1968: three solar eclipses, the Cultural Revolution, and multiple major earthquakes
* 2008-2012: four solar eclipses, a major snowstorm in South China, the Wenchuan earthquake, the global financial crisis, and so on.

The ancient Chinese were very afraid of the occurrence of a total solar eclipse, believing that it would lead to the destruction of the dynasties and the rulers. In ancient times, the emperors issued a "self-reproach edict" and granted amnesty to whole dynasties.

After the total solar eclipse on April 8, 2024, mainland China was affected by the total solar eclipse almost every day until August. Heavy rains and floods continued in the south. But the north was dry. Due to the corruption of Chinese officials, no effective underground drainage system has been established, and the Three Gorges Dam in the upper reaches of the Yangtze River and various dams of all sizes in the middle and lower reaches of the Yangtze River have been opening the gates to release floodwaters without warning, causing half of China to be soaked in water for 4 months. Officials from all over the country were passive in disaster relief, causing serious loss of life and property. This level of disaster can only happen in countries that believe in Satan, the devil, and communism. Does this look like the great flood in the Noah era mentioned in the Bible when God punished mankind? But in the face of cruel facts, there will still be scientists who come out to tell you that the total solar eclipse has nothing to do with natural disasters. This should be because European science lacks knowledge of prehistoric civilization and does not know the true meaning behind the total solar eclipse.

What is the fastest way to learn? The answer is to learn from others' failures to avoid paying a heavy price yourself. However, the current human civilization dominated by European science does not seem to have learned any lessons from prehistoric human civilization. The most obvious evidence is that ancient Chinese science is much better than European science.

Master Li also made a comparison between ancient Chinese science and modern European science. He elaborated:

> *The science of ancient China was different from modern science, which comes from the West. It took a different course and brought about a very different mode of life. It's not appropriate to judge ancient China's science and technology with today's ways of looking at things, for the science of ancient China went a different route, honing in on how the body, life, and universe work together. Meditation and good posture were part of one's education, and when writing, people were mindful of their breathing and directed energy to different areas of the body. And in each of the trades and professions people sought to empty the mind and breathe properly. This was simply the way of things throughout society.*

> *People have questioned whether we would have cars and trains as we do today if we had gone with the science of ancient China; they believe modernization wouldn't have been possible. I would respond that it's not appropriate to judge a different way of life based on one's own, present circumstances; an accurate assessment is only possible with a drastic shift in thinking and concepts. Perhaps in a world without televisions, you would have unimagined powers, such as being able to see whatever you wish through your own forehead. Or in a world without trains and cars, you might have the ability to levitate as you sit, without the help of a lift. Taking a different scientific course would result in the world being a different place, one perhaps quite different from the world around us. The UFOs of extraterrestrial beings travel at unbelievable speeds and have the ability to enlarge or shrink. The course they have taken is even more radically divergent, as it involves an altogether different approach to science." (Zhuan Falun, p318, Li Hongzhi, 1994)*

I would like to explain what Master Li Hongzhi meant by, "*Without a TV, people can wear one in front of their heads and watch whatever they want. This function also exists.*" This is the function that appears when a person's third eye reaches the level of celestial clairvoyance. "*Without trains and cars, people can float up when they sit there, and there is no need for elevators.*" This means that after practicing meditation, the soul can leave the body and go to other dimensional spaces in the universe, including different planets.

Think carefully about how much pollution and damage the current scientific products have brought to our living environment. Take the car, for example, which is an indispensable means of transportation.

A study found that the United States has the highest per capita tire wear, and it is estimated that American tires alone produce about 1.8 million tons of microplastics each year. According to the IUCN, car tires are second only to synthetic textiles and account for 28% of all major microplastics in the ocean. Although the rubber rings under the car look harmless, experts say they are a significant source of air, soil, and water pollution that may affect humans, fish, wildlife, and other creatures. It's just the tires that can cause such a large amount of pollution to the environment, not to mention the fluids in the car that need to be changed frequently.

In the first chapter, we talked about Sun Chulin and how she can go to a space higher than any three-dimensional space, visit different planets, and even visit alien spaceships. If Mr. Elon Musk is really interested in Mars, he should talk to Ms. Sun Chulin. Aliens have shown Chulin the history of Mars from their spaceships. How long will it take for human technology to create a means of transportation that can take us to travel to various planets? No wonder Master Li Hongzhi said that the scientific development of mankind this time has taken a very different, clumsy, and slow path, and in his opinion, a wrong path. It is worth mentioning that the ability possessed by Ms. Sun Chulin is also quite common among many Falun Gong practitioners. They can travel through time and space to communicate with beings in different spaces. Some have seen Jesus, St. Mary, Laozi, Shakyamuni Buddha, and the huge Buddha image reflected by Master Li Hongzhi in different dimensions.

For atheists, immortality is equal to the eternal youth of the body. Brainwashed by human scientific superstition, they take medicine, get injections, and even have blood transfusions to try to prolong the life of the body that imprisons them. Master Li Hongzhi clearly elaborated:

> *When a person dies it is just a matter of his physical body declining and breaking down, while his true soul (which is who he really is, and which does not die with the passing of his physical body) will continue on in a next life, being reborn. So just as the universe goes through Formation, Stasis, Degeneration, and Destruction, so too do human beings go through birth, aging, illness, and death. These are laws of the universe, to which even higher beings are subject, only the time span is longer, with the process being more drawn out in proportion to how great the beings are. Life and death are not painful for them, and they remain cognizant throughout these processes—to them, it is as if but changing outfits. Put differently, normally lives do not really die. " (How Humankind Came To Be, Falundafa.org, Li Hongzhi, 2023)*

III. Why Do We Have Physical Bodies?

Now, with the development of online media. More and more people from different ages, cultural and religious backgrounds, and regions of the world who have had near-death experiences have come forward to tell the truth. When asked what they felt when they were in a state of death, they all talked about a special situation, and they were consistent. At the moment of death, people did not feel afraid. On the contrary, they suddenly felt a sense of liberation, potential excitement, and no pain. Some people felt that they were no longer bound by their bodies, and they floated lightly and beautifully. They also saw their own bodies. Some people also saw life forms in other dimensions, including their deceased relatives, angels, Jesus, Buddhas, Taoists, gods, etc. Some people also went to some places and saw the good and bad things they had done in their lives. Our human body is to make us suffer. In the words of Master Li Hongzhi, "*Everyone comes out of the womb like this, so people can't tell that they are suffering.*" (*Zhuan Falun*, Li Hongzhi, 1994)

I once met a colleague who told me she was blind until she was 20 years old due to diabetes. I asked her sympathetically, "You must have had a very painful life when you were young, right?" She told me that she didn't feel unhappy at all because she had been like that since she was born. She had never seen what the world was like, so she thought it was normal and didn't feel any pain. Most of us are living in this state now, and we don't know what the purpose of coming to the earth is. We are not only attached to this body that makes us suffer, but also to the dazzling array of material products produced by this body, and we can't extricate ourselves. To enjoy these material products, we can harm others to benefit ourselves and do evil.

You are not the physical body that you appear to be, but the soul imprisoned by your body. Master Li Hongzhi also mentioned:

> *There is a Buddhist belief that human beings are fashioned to suffer—that as long as someone is human he must inevitably suffer. They believe that the beings in other dimensions don't have bodies as humans do, and are free of ailments as well as the routine miseries of life; so they aren't subject to the pains that we are. And as beings of other dimensions, they can float about effortlessly and are weightless, which is just wondrous. Whereas normal human beings, saddled with the bodies that they have, are burdened by heat and cold, thirst and hunger, as well as fatigue, on top of which there is the cycle of birth, aging, sickness,*

and death. In a word, it's not pleasant." (Zhuan Falun, p397, Li Hongzhi, 1994)

The human body also has many needs and desires, forcing us to spend a lot of time and life to satisfy them through hard work and suffering. The most important thing is that this body often gets sick and makes us suffer. Birth, aging, illness, and death have become the fate that each of us cannot escape. Therefore, if you want to seek eternal life, you must get rid of this body instead of trying all kinds of weird methods or even black technology to make this body that binds you and makes you suffer continue. One of the main purposes of your body is to punish you. It is through it that you pay off the debt you owe in life, which is called karma, and what Jesus calls sin.

Everything has two sides, and the human body is no exception. The bad side is that the body is used to bind and punish you. The good side is its built with some special abilities. People call them superpowers-just like those of Sadhguru, Sun Chulin, the children with special abilities, and Falun Gong practitioners. These innate superpowers are locked up. This is why the Bible says that God created humans in his own image. It tells you the human body is a downgraded divine body. Humans can also become divine beings, so the miracles God can do, humans can also do, but they need to understand how to unlock these superpowers. I am not qualified to talk about this knowledge, but if you are interested, you can find out more about *Zhuan Falun.*

Some readers may still want to know if there is a way to keep the body young and immortal. The answer is yes. But it is not what you imagine. Once you go that way, you must give up many pleasures in your life. You can no longer live like you are now, and you may even have to be isolated from the world. Most people who pursue immortality are very wealthy people. Their purpose of wanting immortality is to use their wealth to fulfill their dreams and to forever enjoy the material life that wealth brings them. In the following chapters, we will talk about the method of keeping the body young and immortal.

CHAPTER 3:

The Blind Spots of Science–Reincarnation and Karma

In the first chapter, we talked about how Dr. Lee Si-chen could use scientific methods to repeatedly verify the existence of the spiritual world of gods, Buddhas, and Taoists. We also talked about how the evil anti-God regime of the CCP has been studying human body science for many years. Through its research, CCP has learned about the existence of gods, Buddhas, Taoists, and aliens.

In Chapter 2, we talked about scientists who were not as smart as Einstein and did not realize the greatest scientist in the universe was the Creator. We also talked about how European science, which did not inherit the knowledge of prehistoric civilization, had become a kind of scientific superstition. Many people, brainwashed by this scientific superstition, believe immortality means keeping their bodies young forever.

But on the contrary, true eternal life is to get rid of your physical body, because your physical body is used to punish you. In this chapter, we will analyze two blind spots that science does not know–the reincarnation of human life and the karma of human life.

I. Hypnotherapy Proves Reincarnation

For thousands of years, Buddhism and Hinduism have been saying people can reincarnate, but they have been ridiculed as superstitions of Eastern religions. Fortunately, in recent decades, Western psychology has proved the existence of reincarnation through hypnosis.

Dr. Michael Newton, a famous psychologist and hypnotherapist, wrote about the Past Life Regression therapy in his best-selling book *Journey of Souls-Case Studies of Life Between Lives*. His therapy allows hypnotized patients to return to their past life memories and find the relationship between past life experiences and current life illnesses, thereby temporarily solving the patient's current life illness. Delaying or transforming the patient's illness is a temporary solution because the root cause of a person's illness is caused by the karma created in many lifetimes. Karma is a negative energy. And the negative energy of karma is also within the scope of energy conservation. Karma, like other energy, will neither appear nor disappear out of thin air. It will only transform from one to another, or transfer from one person to another, and the total amount of karma can increase or decrease with people's behavior.

A good example here is Ms. Sun Chulin. After curing a five-year-old Japanese boy with cancer, she admitted that she only pushed the child back from five to three years old without cancer in another dimension. But she could not eliminate the boy's cancer. This is why at the Beijing Oriental Health Expo in December 1993, Master Li Hongzhi's students were worried the patients would cause too much pain to the master himself because he took on too much of the negative energy of illness karma for his patients. Christians should now understand why Jesus chose to die slowly on the cross in his own flesh. It was because Jesus transferred the "sins" of too many people onto himself, Buddhists call this karma, but is negative energy. Because Jesus took on too much of this negative energy for too many people, Jesus needed time to slowly transform this negative energy into another form of energy, which was the extreme pain, unbearable torture, and slow death of his flesh. That was why Jesus chose to be crucified. In other words, being nailed to the cross was the only death method giving Him enough time to convert sins from His believers into His own suffering.

Master Li Hongzhi has answered this question:

> *Then what explains all of the energy healing that's being done and taught out there? In most cases, it's that the people involved have gone down the wrong road. True masters of energy practices like chi-gong, by contrast, develop compassion in the course of their spiritual progress and become sensitive to people's suffering. And so it is allowed, when that's the case, for a master to help someone if he feels moved. But what he does won't cure the person. He can only mitigate the ailment for the time being, or push it off to a later date, or he might transfer it to someone else in the person's family. He can't thoroughly dissolve the karma for the person. You can't freely do that for a regular person; it can only be done for people engaged in spiritual practice. So there are principles at work in all of this. (Zhuan Falun, p76, Li Hongzhi, 1994)*

To be more specific, diseases in the human body are the expression of karma, the negative energy, after conversion. But karma can be transferred from one person to another, postponed or converted into other diseases, or even converted into a positive energy called virtue. However, to convert karma into virtue, one must endure hardships, do good deeds, and cultivate. Western hypnotists have not yet discovered this knowledge. Dr. Newton said, "Past Life Regression is still in its infancy." However, the breakthrough of Dr. Newton and his colleagues in hypnosis is a huge leap for Western society and even the entire human race because Western psychology has finally verified the long-standing knowledge of reincarnation and karma in Eastern civilization.

In Richardson, Texas, Dr. Newton's former colleague Mark Johnson owns a clinic called Dallas Past Life Regression. This clinic provides patients with life-between-lives spiritual regression, past life regression, quantum healing hypnosis, and other hypnotic treatments to help patients find the connection between their past life deeds and their current illness and pain and relieve the pain of patients through his expertise. For Falun Dafa practitioners, the current psychology of using hypnosis to make people return to the past and cure patients to a certain extent is only the tip of the iceberg of knowledge about reincarnation and karma.

Megan Pulace, a patient of Mark Johnson's, provided a video testimony for Johnson's website about the effectiveness of her past life regression therapy. Megan went to see Johnson to find answers for some phobias and fears she faced. When mentioning the results of her session with Johnson, Megan reported, "It worked well! I actually have psoriasis, and my psoriasis is starting to clear up due to the session."

In her testimony, Megan mentioned that she was surprised to find that the cause of her psoriasis was not food but a burn in one of her past lives. She went on to say the cause of her death in her previous life was burns. During this hypnosis session, she discovered that she had been burned alive by a man.

During another hypnosis session, she felt herself falling from a great height while crying.

With the help from her hypnotherapist, she soon realized that in this different past life, she was an Indian woman and had done something wrong to anger an Indian man, who was grabbing her by the neck and pushing her off a cliff. She recalled when her body fell off the cliff, the pine trees nearby were clearly visible, and when she landed on a huge rock, her soul rose up and she could see her dead body, with her teeth knocked out of her mouth. With the help of this extremely experienced hypnotist, Megan discovered that her experience of falling off a cliff in her previous life was closely related to her fear of heights and reason for grinding her teeth in her present life.

The reviews of Johnson's clinic are very positive. The patients were initially very skeptical of this non-mainstream treatment method, but after the treatment, they were all conquered by the realistic and effective curative effect. What's more interesting is that patients do not need to go to his clinic to see the hypnotist to participate in the treatment; if they have a computer, they can receive treatment online and the results will be equally good. At the same time, I suggest that hypnotherapists like Mr. Johnson who want to help patients should study *Zhuan Falun* so they can learn about the relationship between disease and karma to avoid relapse of old diseases or transformation of old diseases into new ones.

II. Karma is What Jesus Said about "Sin"

Mother Teresa is a model of human moral life. The so-called model does not mean praising her like an angel but seeing her truth. First, she spent her life helping others, and the whole world could see her love.

However, some people "rewarded" her with hypocrisy, vanity, impure motives, and even conspiracy theories, after Mother Teresa had said, "Pain and suffering have come into your life, but remember pain, sorrow, suffering are but the kiss of Jesus—a sign that you have come so close to Him that He can kiss you." Why did Mother Teresa make such a shocking statement that it has attracted a storm of attacks on her reputation by her opponents and media?

Of course, her statement was shocking to ordinary people. And that's why Mother Teresa has been criticized for allowing the sick to suffer as she believed suffering was a gift from God. Like many people, I once did not understand why Mother Teresa said this. But after I read *Zhuan Falun*, I think I've found the answer.

Karma is the debt you owe for the bad things you have done in your past lives. In 1995, Master Li Hongzhi gave an example when he gave a lecture at the launch ceremony of Zhuan Falun in Beijing:

> *In ordinary people's society, in life after life and generation after generation, people have no idea how many bad things they've done. I would say that people have gotten to where they are today with karma built upon karma—they all have a great deal of karma. It's just like a bottle. If a bottle is filled with dirty things and you twist the cap on really tight, and then you don't pour the dirty things out, when you throw the bottle into the water, "plop," it will sink right to the bottom. If you pour a little bit of it out and you throw it in water, it will float up a little. As you pour a little out it floats a little higher. The more that's poured out, the higher it floats. When you have poured out all of those dirty things and you put the cap on, even if you try to push it down into the water,*

you won't be able to. It just floats there, as that is what its position should be. I gave this example to illustrate the point that people in ordinary people's society have all done some bad things; your heart is polluted, your mind is polluted, and you have a great deal of karma. If you don't get rid of those things and your xinxing doesn't change for the better, you will still produce those bad things. They will keep on being produced every moment. So you must change your heart for the better and get rid of these substances and bad things. In other words, when you are cultivating among ordinary people, you have to suffer a little, your physical body has to labor, and your mind has to bear hardship. Only then can you rise. What I'm talking about is improvement through cultivation. If you don't cultivate your heart, the nature of the universe will restrict you, and it won't let you rise. We can also break it down part by part for you, and tell you how this gong rises.(Teaching the Fa in Beijing at the Zhuan Falun Publication Ceremony Hongzhi Li January 4, 1995 ~ Beijing, China)

Even in human society, if a person does not pay off a debt, they will face the consequences. Therefore, if a person, especially a Christian, does not pay off the karma they owed in a past life, that is, the sins, they will not be able to ascend to heaven and see Jesus. Karma is a kind of negative energy. According to the law of conservation of energy, karma must be converted into pain, suffering, and hardship before it can be paid off. This is why Mother Teresa refused to provide medical help to those patients before the end of their lives.

Master Li Hongzhi once pointed out that the sincere prayer and repentance of Christians, as well as correcting their mistakes, is a kind of cultivation, but Jesus did not tell them because the word "cultivation" did not exist in the language at that time. This kind of cultivation can eliminate karma. Due to the difference in language, the "sin" that Jesus talked about is karma.

Now I will introduce to you a real case of karma repayment that shocked the whole of Thailand to prove the "sin" Jesus talked about is karma.

The most powerful retribution in life that shocked Thailand – Mr. Mek- Winai Kraibutr went from being a Thai superstar to being plagued by a strange disease. There are two reasons why this incident shocked Thailand: first, this Thai movie star is as well-known as Tom Cruise in the U.S. Second, the whole process was recorded, many psychics helped, and miracles happened throughout the process.

The story began in 2017 when Mr. Mek participated in a TV psychic program. At the end of the program, a historic scene occurred. The psychic explained that Mr. Mek was a person who had a clear distinction between love and hate and would never give up and that he must be careful about the following verbal disputes. When the host asked Mr. Mek if he had the personality described by the psychic, Mr. Mek replied, "Yes, I will make those who say bad things about me pay double the price! I will insult and curse those who say bad things about me."

Then the psychic asked, "Have you ever done anything bad to animals?" He replied that since he was 6 years old, he liked to use a slingshot to hunt small animals such as snakes and chameleons. The psychic then asked him, "Do you remember how many animals you have killed in your life? He replied, "No, but a lot; once I went to see a monk in a temple in Ladprao, and the monk told me as soon as he saw me that I had killed a lot of animals in my life." He was surprised because he had never told anyone about this. And he was not famous at that time, so the people there didn't know him. The monk also told him he would have pain all over his body in the future. The psychic then asked him, "Do you have this kind of pain in your body now?" He replied, "Yes, I hurt every day!" The psychic continued to ask, "What kind of pain is it?" He replied, "It's very painful. Sometimes it hurts so much that I can't sleep, sometimes it hurts so much that I can't walk; sometimes it hurts so much that I have to crawl to the toilet; my neck hurts, my back hurts, and my feet hurt all the time."

However, the matter did not end there. Just as the monk in the temple predicted, as time went by, he got a strange disease. Mr. Mek's body also began to change, from a handsome celebrity to a middle-aged sick man with eroded skin. Due to the physical changes, he could no longer stay in the entertainment industry, but Mr. Mek, who had a strong will to survive, began to sell health products he took on social platforms. Three years quickly passed and Mr. Mek, who was suffering from a strange disease, returned to the psychic program. He was overly anxious and wanted to know what would happen to him next. The psychic said bluntly, "I believe you will not be able to recover 100% because your illness is caused by karma." After that, his condition continued to worsen, and he went to 10 famous doctors in a row, but none of them could diagnose the cause. Three years later, Mr. Mek had difficulty walking, so he had to use a wheelchair, and his body became skinny.

Just when Mr. Mek was at a loss for what to do about his illness, Pai San, a well-known local psychic appeared. Pai San usually did psychic readings for others for free, and his psychic readings were of a very high standard. He could see people's past lives and karmic burdens. Under the psychic observation, he learned in Mr. Mek's past life, he was a very powerful person and had offended many people. Therefore, he brought a lot of karma into this life and had to bear it. To solve his strange illness, he had to solve it from the root. Through his psychic ability, Pai San found out about seven debtors who were deeply hurt by Mr. Mek in his past life, and Pai San also knew their names in their past lives and this life, including whether they were reincarnated as humans or animals in this life, and where they lived. Five of the seven creditors had passed away, and Mr. Mek could only confess according to the rituals of the five people's beliefs.

It can be said that the sixth creditor named Jarn Torn is the most difficult to find because there are 7 or 8 people with the same name as him in Bangkok. When Mr. Mek's wife and friends were looking for Mr. Torn, they were told by a former neighbor that Mr. Torn was a civil servant but had moved out of the house ten years ago, and his name had never been on the title deed. With the help of this neighbor, they finally contacted Mr. Torn. Mr. Torn was surprised Mr. Mek found him and decided to visit the sick star and accept his confession and apology. Before meeting, Mr. Torn thought his good deeds would be rewarded, so he bought 10 lottery tickets. As a result Mr. Torn won 9 of the 10 lottery tickets, with the winning amount reaching 180,000 Thai baht, equaling about $5,100. Even more unbelievable is shortly after the reality show was broadcast, local news reported Mr. Torn won the lottery again. This proves the saying, "Good people will be rewarded." I think science cannot explain the series of coincidences which occurred during the process. This incident reminded me of a Falun Gong practitioner who once said to me, "There are no coincidences in this world. Everything is here to test you!"

Just when the sixth creditor finished the sacrifice ceremony, the reality show crew suddenly received a call. The good news was that the seventh and last creditor, had been found. The bad news was the other party was a devout Muslim woman who refused to partake in any confession ceremony Mr. Mek planned for her. Because reincarnation is ridiculous to Muslims. Even though Mr. Mek was a Muslim who grew up in a Muslim family, he also believed in Buddhism. Strangely enough, when the seventh creditor refused Mr. Mek's confession ceremony, Mr. Mek's condition worsened, and he died in pain. It seems that Mr. Mek's fate is like what the psychic said when he participated in the variety show three years prior, "I believe you will never be cured because your condition is caused by karma." Mr. Mek is a Muslim, but he also believes in reincarnation in Buddhism. Because whether it is Buddhism or Christianity, they are all about cosmic science, and human reincarnation is a kind of science, whether you believe it or not. In fact, reincarnation is mentioned in both the New and Old Testaments of the Bible, but it was deleted by the Roman Emperor Constantine and his mother Helena. I remember Master Li Hongzhi said at a lecture in Europe:

> *Religion is a name given by people. When Jesus was alive, he never said that what he taught was Christianity. When Sakyamuni Buddha was alive, he definitely never said that his teaching was called Buddhism.*
>
> *In history, no one has ever told people the true principles of the universe. It has never happened. As you know, Laozi, Sakyamuni, Jesus, etc., only talked about the part of the principles of the universe that they recognized and realized. They only talked about their own understanding, and*

the law of the universe is boundless. (First lecture in the United States; Li Hongzhi, October 5, 1996, in the San Francisco Bay Area)

Many people think some religions are higher-level science. This is why even a smart scientist like Einstein discovered religion is a higher-level science.

After learning about the shocking incident of karma retribution that happened to Mr. Mek, think about why Jesus said everyone has sinned. Think about what Master Li Hongzhi said about karma being the debt owed by the bad things people did in life after life. Let's think about what Mother Teresa said, "Diseases and pain are the kiss of Jesus Christ." I remember Master Li Hongzhi once said Christians in Jesus' time were all cultivators. They improved their xinxing by strictly doing what God wanted them to do and praying and repenting to God devoutly and achieved the purpose of cultivation. But now, most Christians do not regard themselves as cultivators. They turn around and walk out of the church and become ordinary people again because they will fight with others for the interests of ordinary people and even harm others for their own benefit. Therefore, the karma they have accumulated in life after life is very large.

Why do many people now have all kinds of strange chronic diseases? Because these so-called believers do not strictly follow what Jesus said and even distort and misunderstand what Jesus said. Jesus cannot regard them as his believers, but only as ordinary people. So, I think these ordinary people who claim to be Christians can only pay off their karma called "sin" in pain and suffering. On the contrary, almost all Falun Dafa practitioners measure their behavior according to the highest law of the universe Master Li Hongzhi taught them: Truthfulness, Compassion, and Forbearance. They are true believers and true practitioners. In this way, Master Li Hongzhi will bear the karma of these practitioners for them, just like Jesus Christ did in the past. Thus, no matter what kind of serious illness they have had in the past, if they practice sincerely, their illness will soon disappear miraculously. This is why no matter how the CCP persecutes them, they will not give up practicing Falun Gong.

III. Reincarnation of a Christian

Lan Stevenson - a famous scholar who studies reincarnation at the University of Virginia - and Walter Semkiw - a doctor in California who researches reincarnation - shared the following reincarnation case on the "Reincarnation Research" website.

Gamini Jayasena from Sri Lanka was less than three years old when his family took him to Nittambuwa to meet his family from his previous life. The boy's desire to meet his previous family naturally attracted a lot of curiosity and a large group of relatives and friends gathered in front of the house. To test whether he really remembered his past life, his relatives asked him to give a gift to his mother from his previous life. As a result, he easily recognized her and his two brothers among the crowd. So how did Jayasena come to recognize his relatives in the past life?

Jayasena was born in November 1962 in a Buddhist family in Colombo, Sri Lanka. When he was about 18 months old, he began to talk about his past life experience. He said he had another real mother from his past life. He said his mother in his past life was taller than his current mother and had a sewing machine like his current mother. He also said a man named Nimol had bitten him, and Nimol had also been bitten by a dog. Jayasena also had a school bag on a chair at home, which contained books and a toy elephant and he would bathe the toy elephant in the well. At first, the child muttered a few words, and the adults did not take it seriously. When Jayasena was two or three years old, he began to mention his past life frequently and wanted to go back to his past life. At that time, Jayasena's family was very poor and had no electricity, but he knew that the lights could be turned on by a switch. He also said that he wore better clothes in his previous life and he had an Uncle Charlie whose family had a car and a motorcycle. Uncle Charlie would drive him to school.

Jayasena's family also found that he prayed like a Christian, not a Buddhist. He would resist going to Buddhist temples and would not bow to Buddhist monks. When he saw a wooden cross, he would take it home and ask his mother to hang it on the wall. He also liked to celebrate Christmas, talk about Santa Claus, and look forward to Christmas gifts. This made Jayasena's family suspect that he was a Christian in his previous life. In 1965, Jayasena's family took him to Nitambuwa where he pointed to the alley where his old home was and asked to stop the car. He walked into the alley, pointed out his old home, and walked to Uncle Charlie's house. Jayasena's family would not let him enter the house because they wanted to find the truth first and Jayasena cried for 10 days after returning home and begged to return to his previous life's home.

In this way, we saw the scene of "recognition of relatives". The next story is even more interesting. Jayasena's parents in their previous life had three sons. Their family was a Christian family in Nitambuwa. The son, Palitha Senewirat, was born in 1952 and was close to his mother. He had two brothers, one brother named Nimol was 4 years younger than him. Senewirat also liked elephants and he had a toy elephant and would bathe it in the family's well. Senewirat also had a wealthy uncle named Charles who they called Uncle Charlie. Uncle Charlie lived next to his family and had many cars, motorcycles, and electricity at home. These facts were consistent with what Jayasena said. In addition, another detail was even more incredible.

Predicting Death

As Senevirat's family was Christian, he attended a Christian primary school attached to Saint Mary's College. The school was several kilometers away from his home, so he lived in a dormitory near the school. Senevirat seemed to know he was about to die. It was July 1960, and Senevirat was home for vacation and felt unwell. He usually put his school bag and toy elephant in the cupboard, but this time he threw the bag on a chair and said, "I don't have to go to school anymore." Soon, Senevirat's began to vomit. He was then taken to the hospital for treatment but died a few days later at 8 years old.

On the day Jayasena came to recognize his relatives, he found the roof of his previous life's house had become brighter because a corrugated iron roof had been installed. He also found the well where he bathed his toy elephant and knew which person in the previous life photo was himself. Although Jayasena's previous life father and Uncle Charlie were Christians who had no concept of reincarnation, they believed that Jayasena was the reincarnation of Senevirat. His previous life father said, "As Christians, we shouldn't believe in such nonsense, but he recognized each of us, so I think he must be my son, Senevirat, reincarnated."

IV. A Japanese Was Reincarnated as a Cow

There is a very famous monk in Japan named Miki Daiun. He likes to tell supernatural stories to the public. He also wrote “Kyoto Strange Tales,” a compilation of his personal experiences and mysterious stories from encounters accepting commissions. So far, a total of three books have been published. He also opened a channel on YouTube and became an Internet celebrity. However, Miki Daiun said he told supernatural stories, published books, and did self-media not for idleness, but to spread the Dharma to the general public in an easy-to-understand and down-to-earth way, so they could listen to the stories in a relaxed and happy mood, understand the reality of the six reincarnations, and the principle of retribution for good and evil.

Miki Daiun was born in 1972. His father is the abbot of a temple. Some readers may be confused, how can a monk have children? Because Japan’s Buddhist system is quite special, called the “Honmo system”, which divides the temples of various sects into two types: Honsan and Moeji. This Honsan is like the meaning of the base camp. For example, it can be said “the Vatican is the Honsan of Catholicism.” The monks in the Honsan temple are vegan, observe the precepts, and the abbots must be chosen among virtuous monks. The moeji, like the small branches on a big tree, are allowed to develop freely. The monks in the temple are allowed to eat meat and marry. The temple also adopts a hereditary system where the son of the abbot takes over the temple. Miki Daiun is the second son in the family, so his father’s temple was inherited by his brother. In theory, Miki Daiun can learn whatever he wants and does not need to learn Buddhism, but he is naturally interested in Buddhism, and he has been said to have spiritual power since he was a child. How can he have spiritual power? Miki Daiyun said he could smell death since childhood. When a person is sick, he can sense a smell from that person. When the smell becomes stronger and stronger, it means that the person is getting sicker and sicker. When the smell becomes a smell like a ripe orange, the person will have a problem within three days.

Miki Daiyun cited an example of his personal experience. He said one day in a bookstore, he suddenly smelled a strong scent of death. He found the smell came from a person behind him. Out of kindness, he said to the person, "Go to the hospital for a check-up. You may be very sick." The person thanked Miki Daiyun for his concern and said he didn't need to go. Miki Daiyun insisted on persuading the person, but he finally whispered, "Really, no need, because I'm dead." It turned out that Miki Daiyun was talking to the soul of a dead person. Continuing with the smell of death, Miki Daiyun said he had also smelled the smell of the new coronavirus, which was a smell like rust. In fact, Miki Daiyun's statement is scientifically based, because researchers have also found when the human body is sick, it will emit a unique smell. Some animals with a keen sense of smell, such as dogs, are specially trained to find patients in the early stages of the disease by sniffing so the patient can be treated. At present, dogs can help monitor or diagnose cancer, Alzheimer's disease, and malaria, as well as E. coli in newborns. Previously, Germany also reported that dogs were trained to diagnose the new coronavirus with an accuracy rate of 94%, which may be more accurate than the new coronavirus testing reagents in some places.

Witness the Six Paths of Reincarnation

Because of his special family background and special abilities, Miki Daiyun has been on the path of Buddhism and studied Buddhism in college. After college, he began to wander with no fixed place to live and the world as his home. This difficult period made him waver for a while, wondering whether he should continue to cultivate. A friend invited him to study in India where he encountered things which made him determined to be a monk for life. Miki Daiyun said one day he was having dinner with a friend, and an Indian who could speak Japanese came to ask if he was Japanese. This person felt the cow he raised was the reincarnation of a Japanese person, so he wanted to ask Miki Daiyun to help confirm it. This request is quite strange. Miki Daiyun asked, “Why do you think your cow is the reincarnation of a Japanese?” The Indian said he was learning Japanese, and as long as he spoke Japanese, the cow would react. For example, when he says おはよう (good morning), the cow will stare at him, and when he says こんにちは (good afternoon), the cow will nod as if saying hello.

Is it magical? Miki Daiyun followed the Indian to his ranch where he found many cows. Which one could speak Japanese? This was easy. Just say one sentence in Japanese. So Miki Daiyun shouted “Good morning” to the cows. As a result, a cow ran to him and kept nodding. Miki Daiyun then asked in Japanese, “Were you really a Japanese in your previous life?” Unexpectedly, the cow started nodding again, and tears welled up in its eyes. Wow, this is rare first-hand evidence that can prove the six realms of reincarnation. Miki Daiyun was very excited and told the Indian he would go back to the hotel to get a camera to record everything. However, when he came back from the trip, he found the cow was dead. Perhaps the secret of heaven cannot be leaked, and the world cannot be broken; perhaps the cow appeared to enlighten Miki Daiyun, and the mission was complete and it was time to leave. Miki Daiyun was filled with emotion. This cow was a human in its previous life and became a cow in this life. It retained the memory of its previous life, which was an awakening, but also a painful torture. And what about those lives that don’t remember anything and live in the world in a daze? Aren't they also in a kind of suffering? Miki Daiyun decided to practice well, transcend the three realms, and stop reincarnating in the six realms.

V. Officially Recognized Mass Reincarnation Incidents

Previously, we talked about reincarnation cases in the East and the West. Some people were reincarnated as humans and others were reincarnated as animals. Now let's talk about the official cases of reincarnation from animals to humans in Communist China. In Pingyang Township, Tongdao Dong Autonomous County, Hunan Province, many people can clearly remember their past lives and many of them are adults. Locals call these people "reborn people."

Pingyang Township is located at the southernmost end of Tongdao County, at the junction of Hunan and Guangxi provinces. Due to inconvenient transportation, it has little contact with the outside world. The township has a very deep cultural heritage of the Dong ethnic group and simple folk customs. The custom of not picking up lost items on the road and not locking the door at night has been seen there. Every household has some kinship, so people have close contacts, trust each other, and rarely doubt each other. People who have visited feel it to be a paradise.

In 2008, Yang Shengyu, the director of the cultural station of Pingyang Township, opened a blog and serialized 9 stories of reborn people. Each story had a name and was very realistic to read. These stories began to spread quickly on the Intranet of the Communist state. In early 2009, the local news website Huaihua News Network sent a reporter to investigate. The reporter interviewed the then-Party Secretary of Pingyang Township. The secretary introduced: "Although we cannot scientifically investigate the reasons for this phenomenon, this unique cultural phenomenon is very common. Our Pingyang Township has a population of only 7,800 people. We are conducting a cultural survey on this phenomenon of reincarnation, and we have counted more than a hundred people who have been reincarnated." Next, the reporter, with the guidance of the villagers, interviewed two families with "reincarnated people" and confirmed two well-known cases, one of which was the "pig reincarnated as a human" incident.

The reporter interviewed Lu Jutao; the mother of the "reborn" boy named Wu. Jutao said when Wu was a little over one year old, he began to tell her he was a pig, but she didn't think much about it. When Wu was 2 or 3 years old, whenever he saw someone picking pigweed in the field, he would warn them which vegetables were too bitter, which vegetables were too spicy, and if the person picked too much, pigs would not be able to eat it. Moreover, he was always afraid of the village butcher Rong. Every time he saw Rong, Wu would run home. Over time, the villagers felt there must be something suspicious here, so they tried to ask the Wu why. Wu said in his previous life, he was a big white pig raised by his grandfather's family. That day, the butcher Rong brought a man to buy a pig. The white pig saw something was wrong and ran away. It ran all the way to the mountain behind the butcher's house but was caught by Rong and his man and was taken to their house and killed. The story of the white pig's reincarnation as a human quickly spread. From then on, people would simply call Wu "little white pig." Butcher Rong also vowed not to kill again in this life.

Is it hard to imagine all the animals you see are reincarnations of people who did bad things in their previous lives? People always think they are very smart and no one can see them when they do bad things, but they don't know the higher life forms who created us are much smarter than us, and the punishments they use for the bad things we do are beyond our imagination.

After the Huaihua News Network published the report, the "reborn" phenomenon in Pingyang Township went beyond the Dong village and entered the public eye. So, we can't help but ask, how does the local government view this phenomenon? This is not in line with atheism. June 16, 2015, the Beijing News published an interview titled "More than 100 people in Hunan Dong Village claim to be reincarnated and have memories of past lives." The county magistrate, Zhao Xu, told reporters in 2011, Tongdao County officials and relevant experts from the Chinese Academy of Social Sciences conducted an investigation and study on the reborn phenomenon in Pingyang Township. The conclusion was "affirming the existence of the reborn phenomenon, but no scientific basis has been found." Hu Yilong, an official of the Propaganda Department of Tongdao County Party Committee, said he welcomes media reports that seek truth from facts, but does not agree with the voices questioning the "reborn people." He cited an example from 2013, a Jiangsu media concluded in its report that the Pingyang reborn people phenomenon was actually a "collective lie" by the Tongdao government and the public. He said, "We are ready to sue them and hold them accountable for making false news." This means although there is no scientific evidence for the time being, the local government has not included the reborn people phenomenon in the category of feudal superstition and fallacy. I couldn't help but admire them a little. In the authoritarian system of the CCP, there are really not many officials who dare to speak the truth.

So how does the academic community view this phenomenon? A 2015 article from the Beijing News mentioned Huang Jin, a professor of Chinese studies at Central South University, once used a lie detector to evaluate the reborn people. Huang Jin said, "In order to respect the local people's sayings, we deliberately chose the night when the "yin energy" was very heavy to test the reborn people." The Beijing News reporter watched a video of his test. Under the dim light, Huang Jin hypnotized a "reborn person". The teenager fell asleep in under five minutes and Huang Jin began to ask questions, the "reborn person" began to speak in the Dong language. Huang Jin said the lie detector showed the "reborn person" did not lie.

Another scholar, Mr. Li Changzhen, interviewed more than 100 reincarnated people in 2015-2016, and later compiled the interview data and notes into a book "*Pingyang Reborn People: 100 Reincarnation Cases of the Dong Ethnic Group*". The cases included in the book are not only detailed but also supported by more than 400 illustrations, mainly various related photos.

I suggest that folks who like Chinese culture travel to Pingyang Township, Tongdao Dong Autonomous County, Hunan Province, China after the fall of the CCP, and take a few photos with those so-called reborn people as souvenirs.

Master Li Hongzhi mentioned in the *Second Volume of Zhuan Falun* when he was giving a lecture on Lantau Island in Hong Kong:

> *The six realms of reincarnation mean that a person can be reborn as a human in this life and an animal in the next life. In fact, there are more than six realms of reincarnation. This is what Shakyamuni Buddha said. One can be reborn as a human, a*

celestial being, a ghost in the Asura realm, or an animal, a beast, or a material thing.

I heard from a Falun Gong practitioner that reincarnation is mentioned in both the New and Old Testaments of the *Bible*, but it was deleted by the Roman Emperor Constantine and his mother Helena. This is a pity because the *Bible* is the best-selling book in the world, but people cannot learn reincarnation, cause and effect, karma, etc. from the *Bible*. I think the lack of this knowledge has directly led to the decline of human ethics and morality. People now do all kinds of evil things for money and the pursuit of material life.

I learned from reading *Zhuan Falun* that people on Earth are very complex and come from different times and spaces, different dimensions, and different planets. All people have done bad things at their level and have been falling. Finally, they fell to Earth and suffered in continuous reincarnation to eliminate the karma of the sins they had committed. So I think this is why some people suspect Earth is a prison or a labor camp. Do you now understand why Jesus said that everyone is sinful?

Any being that falls to this plane should have been destroyed or annihilated. Yet they have been spared by higher divinities who created this special dimension and human world, out of a great sense of compassion. Here, beings are given an additional, mortal body, along with a pair of eyes that are limited to seeing only this material world. So it's as if these beings are put under a spell of ignorance that prevents them from seeing the universe as it truly is—which is not the case in any other dimension. By being put into this state of ignorance, they are given an opportunity, however. It is extremely painful here, while lost and confused, and the human body is fashioned to suffer. Yet if a person in this setting wants to find his way back to his true, heavenly home, or return to his original state of purity, as Daoists put it, if he wants to dedicate himself to spiritual practice, it means that his innate divinity has shone through, and higher beings will help him, for they value that intention above all others. It would mean that this person hasn't lost his way, even in such a trying setting, since he wants to go back. And so higher beings will help him unconditionally and in every way possible. And so this is why we can do so much for those who practice, but not others. (Zhuan Falun, p74, Li Hongzhi, 1994)

Master Li Hongzhi pointed out directly in his article *How Humankind Came To Be* published in 2023:

This is to say that the purpose of people's lives on this earth isn't to accomplish something in the world. All of the intense efforts and attempts people make in life, and their drive to get what they want, which can even involve resorting to unscrupulous means, only make people immoral in the end. The reason people came to this world and became human was to atone for their sins and karma and to make significant spiritual progress. People came to this world to gain salvation. They came and assumed human form to await the Creator and His salvation back to their heavenly kingdom. And while they waited, they built up merit over their many past lives; and that was the purpose of people's rebirths. The troubled nature of this world is meant to make something great of these lives. Of course, there are some people who, when seeking divine help in times of duress, haven't been satisfied with the outcome, and started to loathe God—even turning against Him as a result. Some have even turned to the demonic, dark side, and committed still further sins and made yet more karma. Those whom this applies to had best quickly come around and beg God for forgiveness if they are still to have a chance of reaching safety. Everything that happens in one's life—whether it seems warranted or not—is, in reality, the karmic consequence of what one did in one's past lives, for better or for worse. The amount of blessings and virtues that one built up in one's past lives determines what fortune is in store in this lifetime, or perhaps the next. If one lives a blessed and virtuous life now, perhaps it will translate in one's next lifetime into a high position and salary, or it might translate into different kinds of wealth and fortune. This would also include whether one has a happy family, or even how one's children turn out, and so on. This is the fundamental reason why some people are wealthy and others poor, why some hold positions of high rank while others are destitute and homeless. It's nothing like the diabolical nonsense that sinister communism spouts about equality between rich and poor. The universe is fair. Those who do good are blessed for it, while those who do wrong will face payback—if not in this life, then in the next. For this is an immutable law of the universe! Heaven, Earth, the Divine, and the Creator alike are compassionate toward all lives. Heaven and Earth, just as with man and the divine, were made by the Creator, and it is never the case that He plays favorites with some lives and shortchanges others. The reason some people lead happy lives and others do not all comes down to rewards and retributions for past deeds.

I think this is why you should believe in God as early as possible, do more good deeds, create less karma, work hard, and help others. Because if you don't do this, the old karma you created in your past lives and the new karma you created in this life may soon exceed the threshold. You will suffer retribution in this life, just like the movie star in Thailand. What's worse is if your karma accumulates too much, you may be reborn as an animal next time, or it may be transferred to your family members.

> *In our observation, the deeper reason people fall ill or experience misfortune is karma—a field of black matter. It is negative, or yin, by nature and something bad. Evil entities are yin natured as well, and dark, and they can enter a person's body when conditions suit them. And they are the root cause of the ailments that people suffer, the main source of illness. But there are also two other ways that illness comes about. One involves extremely small, negative entities that are high in density and resemble clumps of karma. The other, which is less common but does exist, is where it is piped in from one's ancestors. (Zhuan Falun, p307, Li Hongzhi, 1994)*

But if you know the truth of the universe and act according to the highest laws of the universe, then you have a great chance of getting rid of this flesh that binds you, achieving immortality, and returning to your previous dimension or going to a higher dimension. In the next chapter, we will discuss specific methods of getting rid of the flesh and achieving immortality.

VI. Reincarnation and Karma Scientifically Explained

Here I quote Master Li Hongzhi's explanation of reincarnation and karma from the perspective of contemporary science. I hope it can help readers better understand that reincarnation and karma are real.

But consider this. I just spoke about karma and virtue. Karma and virtue are carried along as humans reincarnate. A human can bring along nothing from his life except for these things, which follow him through the cycle of reincarnation. Speaking of reincarnation, I want to tell you that religions talk about humans going to other dimensions after death; particularly in the religions of the East, they talk about the sixfold path of reincarnation. Humans can indeed reincarnate. This is true. There is no doubt about this among cultivators, and the concept is well understood. Why is there reincarnation? There are people who say, "When people die, that's it, isn't it?" What dies is the part that grows from eating human food after birth; the human life itself is not dead.

To put it in a way that modern people can understand, think of it as follows. The human body is composed of surface-layer materials that are comprised of molecular particles. This much is familiar to all. The air surrounding the planet Earth, wood, and cement, iron, and steel in buildings are all surface materials that are comprised of different molecular particles. Then molecules are made up of atoms, and atoms are made up of neutrons, electrons, and nuclei; and going further down from nuclei, quarks make up nuclei, and nuclei are made up of neutrinos. If you try to trace further down, humans don't know what is there. When humans die, only the flesh in the surface material dimension that is comprised of molecular particles is shed; it's like the person's clothes are taken off. And the body parts made up of atoms, nuclei, and materials smaller than quarks are not dead in the least. They couldn't possibly die along with the surface body. Think about it: when a nucleus is split, a nuclear explosion will result. When a human dies, could there be a force strong enough to cause nuclei to split? When today's science intends to split nuclei, high heat, and forceful impact are needed to facilitate the change. How could an ordinary human body have strong enough energy to bring about that fission? The temperature of a crematory is definitely not high enough to cause nuclei to split. In other words, the more microcosmic substances that make up the human body cannot be destroyed by the fire in a cremation furnace. If it could really split the nuclei in your body, there would be a nuclear explosion. When the nuclear materials within the human body exploded, it would cause the destruction of a large city. It is an atom, and its energy is enormous. So why has this never happened? This tells us that the atomic elements within the human body have not been destroyed.

As you know, nuclei and atoms are highly radioactive as far as humans are concerned. In other words, they are a kind of energy. In fact, though, substances still more microscopic, such as quarks, have even stronger radioactivity than tnuclei, and no one knows how many times stronger the radiation of neutrinos is than that of quarks. The smaller the particle, the stronger its energy. I can say to you that the gong that you cultivate contains strong atoms, neutrons, and even more microscopic substances. Why can the gong that you develop heal illnesses? Why can it change the human body? Why can cultivators perform so many miracles? It is because gong and supernormal abilities are composed of these higher-energy substances. Moreover, these substances, which are developed through cultivation under a true Law, are alive and have a kind nature, and they are controlled by the cultivator's main consciousness and directed by his mind. They are unlike the malicious, destructive effects that result from the nuclear fission that is done using scientific means. The force released from the explosion of an atomic bomb using scientific means is malicious. It doesn't go in a specific direction, it is harmful to human beings and other lives, and it is destructive to the human living environment. The energy emitted from cultivators, by contrast, is conscious and can have a positive effect. Researchers from the Chinese Academy of Sciences once conducted tests on me to come up with an appraisal, and they could detect energy matter emitted when I was teaching the Fa. The Institute of High Energy Physics of the Chinese Academy of Sciences specializes in high-energy physics research. They placed instruments in the conference room's four corners, as well as in various spots in the middle of the room. An instrument was also placed on the table behind where I was teaching the Fa. During the test, they found that the energies I emitted contained strong neutrons as well as atoms. Of course, that was as far as they could go in their testing, as they did not have instruments for more microscopic substances. They were really surprised that the energy fields I emitted were evenly distributed and the energy could be directed, because scientific research says that nuclear matter doesn't have a specific direction when it is emitted and no one knows where it will go, and it's stronger closer to the source of emission and weaker as it gets further away. And the radiation is supposed to be harmful to everything within its range. Clearly, modern science is not perfect, and blindly believing in it will do extreme harm to humanity.

I just explained that human life does not die along with the physical body, so the soul of a human being will be freed as the flesh body in this dimension dies. It was in a different dimension originally, and only merged with the human body in this dimension at the time of birth. When the body in this dimension dies, the soul is freed. Thus, the cycle of rebirth that some religions have spoken of is real. A life can reincarnate in the cycle of rebirth. A person can reincarnate repeatedly as a human being, or reincarnate as an object, as an animal, or even as a higher being, or as something else.

Buddhism talks about the unlocking of the Five Powers, among which are Celestial Vision, Wisdom Vision, Law Vision, and Buddha Vision. When your Law Vision is unlocked, the world before your eyes will not be like [it is here]. What will it be like, then? You will find that your eyes will be able to, within your level, see through any object and observe even more microscopic substances. And you will find that all objects are alive. When these objects realize that you can see them, they will communicate with you, using language and thought to communicate. Some ordinary people might feel that this is too good to be true, and absurd. Of course, these are matters of cultivation, and they are being taught to students of Falun Dafa. You are all cultivators, and I am not talking to ordinary people; I couldn't just casually tell ordinary people these things. Many ordinary people wouldn't believe them.

So, by then you will have found that the life of an object could have been a human being before; he had died and reincarnated as an object. At this point I want to say something. A human being carries karma as well as virtue along with him while reincarnating. As the morality of today's people declines, people have less and less virtue and ever larger amounts of karma. Because of how lives reincarnate into the forms of each other, when you now take a look, you will see that not only human beings but also objects have black karma. A life will carry this karma as it reincarnates in the cycle of rebirth, so any object might have karma, which could make someone fall ill. In the human dimension karma manifests as microscopic viruses. Nowadays the karma is so abundant that anything might carry it. As you may know, in the past when a Chinese farmer got a cut on his hand while working in the fields, he would grab some soil and apply it to the wound, giving it no further attention. And soon the wound would heal. But today you'd better not even touch dirt [when cut]. With even the slightest touch an ordinary person might be infected, making the wound fester, and he might even die of tetanus. Why is this? This shows that the soil nowadays is soaked with karma. So, when one looks at the Earth from a higher dimension, karma is seen everywhere—waves upon waves of it. Since the human eye can't see it people think that things are all right.

Everyone knows about the flu, right? Basically, the flu breaks out when a cluster of high-density karma rolls over an area. Cancer, AIDS, and the like are malignant karmic diseases that have specific targets—for example, AIDS targets promiscuity and homosexual behavior—and they are karma of even higher density. Generally speaking, in places with more karma people will get sick. Epidemics will break out in whatever region has high-density karma, and that place is afflicted because of the large amount of karma created by human beings.

Why am I talking about this? I have seen that human morality has declined drastically and that if this continues, human beings will face even greater danger. If gods in the heavens don't regard human beings as human, then real danger is imminent. Humans should act like humans—having a human form doesn't necessarily make for a human being. Monkeys have forms similar to humans, after all. If human beings don't have human moral norms and standards, gods will not call them human, and in that case human beings are in grave danger. That's because human beings are created as well as governed by gods. Mankind's authentic cultures appeared when gods willed it to be so. When human beings deviate from human norms, gods eliminate them. I have found that as human beings drift along with the current of things, they have unknowingly made everything worse, become worse themselves, and caused society to deteriorate. And most notably, they have fouled up human morality. The most obvious manifestation now is how culture has turned demonic, leading human nature to turn wicked. As a result, organized crime, drug abuse, drug trafficking, promiscuity, and homosexuality have cropped up in society. People are drowning in a sea of lies and countless vices. Even criminal kingpins are glorified nowadays. There are many, many things like this. Think about it, everyone: is this normal? How human concepts have changed! And it's not limited to these issues. Strong demon nature now exists in the minds of people from all different segments of society. I am teaching this Fa for the purpose of freeing human beings from this danger and enabling them to be saved through cultivation. If you can truly cultivate to the end, I really can make it possible for you to achieve Consummation.

Just now I mentioned some phenomena in society. At present, it's not that I want to do something for society; I haven't had that thought. However, this Fa is able to save people, it teaches people to be good, and it can truly change your xinxing and fundamental nature. Thus, although there are many people who don't do cultivation, when people know the teachings of this Fa they will try to be good people. When people realize what harm they will sustain if they become bad (applause), they will try to be good people. This means that when a true Fa is made public, it will surely benefit society. (Teaching the Fa at the Founding Ceremony of the Singaporean Falun Dafa Association Li Hongzhi July 28, 1996 ~ Singapore)

VII. The Significance of Reincarnation Knowledge

If you call America home, you must learn about reincarnation because we are a country irreconcilably divided.

The United States is a country built on Christian civilization, and Christianity is the world's largest religion. Both Buddhism and Hinduism know about reincarnation, but Christianity, the largest religion, does not know about this, which is very abnormal.

That's why I think the Roman Emperor Constantine and his mother Helena's removal of stories about reincarnation from the Old and New Testaments was a form of cancel culture against Christianity. Just like the Cultural Revolution launched against China more than 70 years ago, and the cancel culture against conservative values in the United States today. The only difference is that in China they used term "CLASS" is to make the poor fight the rich and in the U.S., they use the term "RACE" to make a different racial group of Americans fight each other.

Those evil lives really have a long history of hatred Against Humanity. I think that's why they took the stories about reincarnation out of the Bible because they wanted to cover it up so they could stir the racial pot in America. Think about it, if I hated people of a certain race very much, but after hypnosis of past life regression I went back to one of my past lives and found I had been someone of the race I hated the most in this life so, what would be the point of hating each other?

I once heard a story of a woman who had a near-death experience. She was a man who had killed a woman in her previous life. According to her, during this life, the doer (killer) became the receiver (victim); their roles got reversed.

Our skin colors are nothing but outfits. The real you is your soul under the outfits of skin color, not the outfits of skin color themselves. We are constantly wearing different outfits in reincarnation. Why should we hate each other for a set of outfits?

CHAPTER 4:

Psychology Proves There Are Ways to Achieve Immortality

In the first chapter, we talked about the Taiwanese scientist who accomplished what Newton had failed to by using scientific methods to repeatedly prove the existence of God, other divine beings, and the spiritual world. We also revealed the Chinese Communist regime, the world's largest axis of evil, also discovered the existence of gods, Buddhas, Taoists, and aliens when studying human body science.

In Chapter 2, we talked about how scientists who were not as smart as Einstein did not realize the greatest scientist is the One who made us. We also talked about how European science, which did not inherit the knowledge of prehistoric civilizations, had become a kind of scientific superstition. Many people, brainwashed by this scientific superstition, believe immortality means keeping their bodies young forever.

In Chapter 3, we revealed the scientific blind spots of this cycle of human civilization, namely reincarnation and karma, and their impact on our lives and immortality.

In this chapter, we will talk about the specific methods which can enable us to obtain true eternal life and the secret of what is so-called "physical body immortality"

I. The European Psychologist Deciphered the Ancient Chinese Book of Immortality

The first Westerner to crack the Eastern way to immortality was the famous psychologist, Carl Jung. This great psychologist wrote a psychological masterpiece called *The Secret of the Golden Flower- A Chinese Book of Life.* This book has been very popular in the Western world since the beginning of the last century and was finally translated into various languages and became popular all over the world. The meditation culture that has emerged in the West in recent decades is also closely related to this book.

Here, I would like to give you a brief introduction to the origin of the psychology work *The Secret of the Golden Flower* and Jung's personal understanding of the ancient Chinese way of becoming immortal as a Western scholar.

At the beginning of the last century, when a large amount of Western culture poured into China, some of the essence of Chinese traditional culture was also introduced to the West by some discerning blue-eyed China experts. In terms of inner alchemy, two classics, namely *Taiyi Jinhua Zongzhi* and *Hui Ming Jing,* were translated into German and attracted the attention of some knowledgeable people in the Western world. With the recognition of Jung, the founder of the modern Western school of analytical psychology, they gradually became widely circulated and became one of the strong pieces of evidence of Jung's "collective unconscious" category.

When talking about the translation of *Taiyi Jinhua Zongzhi,* we must briefly introduce its translator–

Richard Wilhelm, the great contributor to the spread of inner alchemy to the West.

Richard Wilhelm was born in Germany in May 1873. He graduated from the Department of Theology at the University of Tübingen in 1895. In 1897, he became the associate pastor of the Prudistant Church and in 1899, he went to China and became the pastor of the Qingdao Church in Jiaozhou Bay, which was then a German colony. He lived in China for 21 years. Jung once recalled that he was "different from ordinary pastors. He was deeply attracted by the charm of Chinese culture. At that time, Chinese intellectuals admired Western civilization and ignored China's inherent culture." But he "devoted himself to studying Confucianism and Taoism abandoned by the Chinese." He loved China so much that his "writing posture, way of speaking and appearance were completely like a Chinese." It is rare to find such a Westerner who loves and understands China. We will never forget him and his contribution to the spread of Chinese culture to the West. Wilhelm came to Germany in 1920 and was appointed by the German government as academic adviser to the German embassy in Beijing in 1922 where he returned to China. He went back to Germany in 1924 and later lectured on Chinese studies at the University of Frankfurt, served as director of the Institute of Chinese Studies, and edited the *Journal of Chinese Studies and Arts*. He died of illness in Tübingen in March 1930.

During his stay in Qingdao, Wilhelm learned that Mount Laoshan was one of the birthplaces of Taoism in China, so he went to Mount Laoshan to preach Christian doctrines to the Taoist priests in the local Taoist temples, because he believed that Taoism in China was just some superstition practiced by the local indigenous people, not an orthodox religion like Christianity.

After explaining the Christian doctrine to an old Taoist priest on Mount Laoshan, the priest said, "Since you are here for cultural exchange, I will also show you some of our Taoist knowledge". Later, the priest showed him some supernatural powers which can only be seen in contemporary science fiction movies, that was, the skills cultivated by Taoist practitioners. After witnessing these supernatural powers he was shocked and begged the Taoist priest to let him learn more about Taoist practice. I think the supernatural powers the priest showed to the German missionary must be far superior to those of Ms. Sun Chulin.

The two men established a friendship, and the priest generously gave him a Taoist secret book on cultivating immortality, *Taiyi Jinhua Zongzhi*, which could only be spread in a small circle of Taoists. Wilhelm sent the translated version of this book to his good friend, psychologist Carl Jung, who was in a career downturn in Switzerland. After reading it, Jung praised it and called it a divine book; and claimed the book helped him solve problems while studying the collective unconscious of human beings. Jung also wrote *The Secret of the Golden Flower, the Chinese Book of Life*. As he said in the book, the ancient Chinese revealed the truth of the world with simple language, bringing the elegance and fragrance of the golden flower. It gave Westerners a new understanding of life and Tao. This book is a classic example of the collision of Eastern and Western cultures. It is rare work by Western scholars to explain Taoist cultivation from a psychological perspective. Once published, it caused a sensation throughout Europe and became one of the best-selling books in Europe. Later, it was translated into English and Japanese where it became extremely popular, especially among those practitioners and readers who like to explore the truth about the universe and life.

The reason why it caused a great response in the West is not only because of its unique content and the charm of being a foreign culture compared to Western culture but also closely related to the introduction and annotations of Richard Wilhelm. First, he viewed the social background of the popularity of *Taiyi Jinhua Zongzhi* with a calm and objective attitude toward Westerners. He said:

> *A series of mysterious sects have been formed in China. They strive to practice the ancient mysterious traditions to achieve a certain spiritual state and escape the suffering of life. Their methods of practice include drawing talismans, praying offering sacrifices, etc. In addition, there is also witchcraft such as shaman dancing that is widely circulated in China. This is a means to establish direct contact with gods and the dead.*

Taiyi Jinhua Zongzhi is a Taoist practice scripture written by Taoist master Lu Dongbin, the Pure Yang Zhenren. The book focuses on the practice of the Taoist inner elixir (Jinhua) and uses Confucian and Buddhist books to explain it, so the text is relatively easy to understand. The main purpose of *Taiyi Jinhua Zongzhi* is to refine essence into qi. The book follows the Taoist inherent "semen retention" theory and further develops it, advocating that "essence" (semen) or "blood" must first be transformed into "qi" and circulated in the body to make the immortal pill of the flesh. The "Taiyi" in the title refers to the original fresh and perfect state of the universe.

Taiyi Jinhua Zongzhi involves meditation and qi circulation, advocating breath retention and reducing conscious activities to avoid "leaking consciousness". It believes there are two invisible things in people, "primordial spirit" and "consciousness spirit". The primordial spirit is part of the illusory world. It is difficult for ordinary people to perceive its existence and function. The consciousness spirit is the physical feeling and the seven emotions and six desires driven by the body. Most people are trapped in the realm of the consciousness spirit throughout their lives and rush to satisfy the consciousness spirit, but it is meaningless. Sometimes people can unconsciously drive the function of the primordial spirit in dreams and start to control the universe. In that realm, the mysteries of the world and everything in the universe can be seen at a glance. However, as soon as they wake up, they forget everything or only remember unclear fragments. This is the life of ordinary people. However, the most important thing in life for those who practice this method is to deeply understand and develop the function of the primordial spirit. Through inner practice skills, it can achieve complete suppression and expel all conscious feelings. The whole person becomes a pure yang body occupied by the primordial spirit, so it is called pure yang learning. That realm is to become immortal. The immortal is to be able to drive the primordial spirit at will with thoughts, to directly operate the cosmic origin of the empire, accomplish things humans cannot do, and breakthrough time to know the past and the future.

In the German version of Richard Wilhelm's *The Secret of the Golden Flower*, one of the contents taught to him by his Chinese teacher is that those who have successfully practiced *Xinggong* or *Xinxing* refer to the comprehension and cultivation of the mind and character, can see wonderful, shining patterns appearing in front of their eyes. Wilhelm called this wonderful light pattern a mandala. Europeans first learned about mandalas from Buddhist scriptures. Some Buddhist scriptures, such as the *Mahāvairochana Sutra* and the *Vajraśekhara Sūtra*, borrowed this word to describe the special phenomenon that occurs during *Xinggong* or *Xinxing* practice. Chinese Taoists believe this phenomenon is that a bright, wonderful pattern will appear at the Tianmu point, which is the third eye, in front of the practitioner›s eyes. The pattern seen in practice was drawn down and called a Danqing Mandala. What the world has seen are just some Danqing Mandalas. Only those who have practiced *Xinggong* or *Xinxing* to a certain degree can see their own mandalas. This is the real mandala. This pattern can reflect the information induction between oneself and the universe and life. In *The Secret of the Golden Flower,* Wilhelm gave some mandala patterns he collected. The mandala seen by practitioners is the light of human nature or the light of the primordial spirit. Refining the mandala means people begin to see their own light of nature, which is called «*the beginning of the return of light*" in the book. The change of light of nature from emanation to return is a very important and meaningful phenomenon in this practice process, and it is proof of entry. The book says that the return of light has different levels "Golden Flower Suddenly Spitting", "Golden Flower Rightly Blooming", and "Golden Flower Great Condensation". When you reach the state of "Golden Flower Great Condensation", the primordial spirit that has been motionless will move. Once the primordial spirit moves, you can get a glimpse of the reality of life.

In fact, Westerners, especially those rich people, like to meditate, which is also closely related to *The Secret of the Golden Flower.* Here, I want to tell you, whether it is Wilhelm or Jung, from the perspective of Westerners, they could not fully understand the content of the book of *Taiyi Jinhua Zongzhi.* This cannot be blamed on them because as European scholars who had not accepted any knowledge from the last prehistoric civilization of mankind, they were able to write *The Secret of the Golden Flower* to introduce the mysterious and ancient Eastern

Immortality technique to Western readers. It is already very remarkable, but I want to emphasize this is also why practitioners who practice meditation through this book wrote by them can only use their third eye to see different "golden flowers" rolling constantly, but cannot see the truth of other dimensions of the universe and the existence of other advanced life forms; not to mention transcending life and death through the methods introduced in the Chinese book of *Taiyi Jinhua Zongzhi.*

I advise those who have read *The Secret of the Golden Flower* to read *Zhuan Falun* because German sinologist Wilhelm and psychology master Jung only understood ten percent of the book *Taiyi Jinhua Zongzhi.* If you practice according to their translated and interpreted version of *Taiyi Jinhua Zongzhi*, that is, *The Secret of the Golden Flower*, at most you can only see the golden flower and gain some wisdom during meditation, but it cannot help you transcend life and death. Because they didn't understand *Taiyi Jinhua Zongzhi* is a Taoist secret book for practicing becoming immortals and is not casually given to ordinary people.

Taoists in China all know when a master is passing on his true knowledge to his disciples, he will select several disciples but will only pass it on to the one who has passed the test. This must be the disciple with the best moral character and the best mind because only this kind of disciple has less karma and is more likely to become immortal. Master Li Hongzhi has a clear explanation of this in *Zhuan Falun*:

> *Let's turn our attention to virtue now, and see how it's related to what we have been discussing so far. I'll explain it in some detail. Each individual has multiple bodies that exist across many dimensions. The largest component of the body*

that the eyes can see is the cell, and cells make up the physical body. But if you were to enter into the space between cells and molecules, or one molecule and another, you would experience what it's like to enter another dimension. To go there, however, your body would have to first meet the requirements for existence there. But the concepts we have here wouldn't enable you to understand the very different bodies you saw there—bodies that can expand or shrink. And those are unimaginably vast realms. But I am just describing one simple form of parallel dimension, which exists in the same place and at the same time as ours. A person has a different body in each of various dimensions, and in one particular dimension, he has a body that is surrounded by a field filled with virtue. Virtue is actually a white material and not just a moral or theoretical concept. It is very much physical. And this would explain the old sayings in China about "accumulating" or "losing" virtue. There was really something to them. Virtue surrounds a person's body and forms a field. And so in the Daoist tradition, the common practice is for masters to seek out disciples, rather than vice versa, since they can see how much virtue someone's body has and what his spiritual prospects thus are. Someone with much virtue would have good prospects, while someone with little would not, and it would be hard for him to develop higher energy. (Zhuan Falun, p32-33, Li Hongzhi, 1994)

What Wilhelm didn't know was that his Taoist master gave him *Taiyi Jinhua Zongzhi* because he had more virtue, which might be related to the fact that he was a devout Christian pastor. Although the great psychologist Jung explained the Taoist immortal cultivation method from a psychological perspective, due to the differences between Eastern and Western cultures, he did not know "virtue", a substance invisible to the naked eye, plays a decisive role in immortal cultivation. This is why most Western practitioners can only see some strange lights and mandalas with their third eyes when they meditate according to the method explained in *The Secret of the Golden Flower*. When this state occurs, it means the meditator's third eye has been opened, but because the practitioner does not pay attention to improving his or her personal morality, the negative energy of karma the practitioner brings cannot be converted into positive energy such as virtue. So the person has been in this situation where he or she can only see mandalas for a long time and cannot sublimate to a higher dimension to see the truth and life forms of other levels. The meditator is even more unable to escape from his or her own body of imprisonment and reach the realm of immortality.

In *Zhuan Falun*, there is this explanation of the mandala practitioners see when they reach the initial stage:

> *It's rather challenging for a person to open the inner eye on his own, by contrast. Let's talk about the ways that it happens. One way involves gazing at the forehead, where the inner eye is while meditating. At first, the forehead will be dark for someone who does this, with nothing there. With time, however, the color will gradually become less dark. As more time goes into the practice, the forehead gradually brightens until ultimately it takes on a red glow. At this time it will seem as if the forehead is blossoming, with a scene similar to how flower buds open almost instantly in nature films. The red part of the forehead, which originally appeared flat, will bulge suddenly in the center and bloom, again and again. If a person tries to work through this*

> *process all on his own, even eight or ten years might not be enough, for his entire inner eye would have started out being completely blocked up. (Zhuan Falun, p63, Li Hongzhi, 1994)*

If you like to sit in meditation and have read *The Secret of the Golden Flower,* which explains the Taoist cultivation of immortality with psychological principles, then I suggest you read *Zhuan Falun* because it is a guide for you to cultivate immortality. If you strictly follow the methods introduced in this book, it can help you quickly clear the karma in your body, that is, get rid of all diseases and achieve eternal life. In fact, this book is a guide for all believers of orthodox religions. Another benefit I gained from reading this book is the questions in the *Bible* which have troubled me for a long time have been answered one by one.

We know the *Bible* is a book about prophecy, the origin of humanity, and human history but it is also a collection of disjointed, incoherent words people recall as being said by God. We know the people who wrote the *Bible* were from different cultural backgrounds in different historical times. People make mistakes, right? So, it is normal there will be errors in translations and interpretations. In addition, the time, place, and situation of the words that Yahweh and Jesus spoke to people have changed. If we think logically, we know the people who listened to Jesus' teachings were shepherds, farmers, and small handicraftsmen, and many of them were illiterate. They were not like most of us today who at least graduated from high school and many of you received higher education. This is why some pastors cannot understand the *Bible* and instead, interpret the *Bible* at random. *Zhuan Falun* has a very strong reference function, especially for some difficult-to-understand and confusing scriptures in the *Bible.*

Falun Gong or Falun Dafa, this miraculous practice that leads to eternal life, has gone through 25 years of brutal persecution and more than 30 years of severe trials and has been witnessed by hundreds of millions of people. It was initially popular only in China, but now, despite the brutal persecution by the CCP, it has grown from being only in China to now having free Falun Gong volunteer counseling centers in every city in more than 150 countries. Curious people can easily find the contact information and address of their local 100% free volunteer counseling centers on the Falun Dafa website Falundafa.org. It is worth noting these volunteer Falun Gong counseling centers have no membership fees, no one asks for your donations, and no hidden fees. If anyone asks you for any fees, it must be a fake volunteer counseling center set up by the local CCP agents in your country to slander Falun Dafa.

II. Physical Rejuvenation and Eternal Youth

In Chapter 2, we talked about the possibility that some readers still want to know if there is a way to keep the body young and immortal. The answer is yes! It is not what you imagine because once you go that way, you must give up many pleasures in your life, you can no longer live like you are now, and you may even have to be isolated from the world. Most people who pursue immortality are very wealthy people. Their purpose of wanting immortality is to forever enjoy the wealth they and their ancestors have accumulated, which seems to be endless. Here I would like to share my limited understanding of keeping the body young after reading *Zhuan Falun.*

You may have heard in India many yoga practitioners can completely sink their bodies to the bottom of the river or bury themselves in the ground for days or even weeks. In Nepal and the Himalayas, many practitioners can sit in the jungle in a meditative state without eating or drinking for months or even years. Their Chinese counterparts prefer to meditate in famous mountains and rivers, and some of them can meditate for hundreds or even thousands of years. The actual human age of these practitioners is hundreds or even thousands of years. There are even practitioners whose physical age has reached more than 6,000 years, which is longer than the time of the last Great Flood. Of course, you can't see it, and you won't believe it, because an ordinary person's naked eyes can't see the aliens disguised as humans walking on the street.

Speaking of aliens, I believe people remember the three unauthorized and released unidentified aerial phenomena (UAP) videos recorded by Navy aircraft carrier pilots in 2004, 2014, and 2015, which have been widely reported by the media since 2017. Later, under pressure from public opinion, the Pentagon officially released these three unidentified aerial phenomena (UAP) videos in 2020 and confirmed the source of the leaked videos in two statements in 2021. More truth about aliens will be described in the answer section of this book. Is it more unbelievable for you to believe there are these practitioners who can live for hundreds or even thousands of years in the world than to believe in the long-standing alien life in the world?

Why are these things unbelievable to the current human beings? Because the human brain has been brainwashed by scientific superstition. In communist countries, this is called the government's brain control of the people. This is also why those practitioners must isolate themselves from the world and go to the mountains and forests to practice. In fact, they are clearing all the acquired concepts in their minds. These acquired concepts are instilled into them by imperfect humans. Human concepts are not the truth of the universe. Everyone is created by God, so these people are very clear they need to get rid of human concepts first. When born, everyone's mind is like a huge, blank whiteboard. In the growth experience of life, good people, bad people, people who think they are great, people with ulterior motives, and even the devil began to write down the definitions and lies they think are the "truth" on this whiteboard.

After practicing the definitions written by all the people, the practitioners painfully discovered following the definitions written by these people could not solve the problems in their lives, especially the biggest problems in life–why did they come into this world? Why did some things happen to them? And where will they go after they die? So, after they were at their wits' end, they thought of God when they were most hopeless. Then they began to understand religion and theology and were surprised to find many of the problems in life had already been answered in some orthodox religions. At the same time, they understood people came to this prison on Earth because they had done too many bad things in the past and owed too many debts, which is karma (sin), so we were constantly reincarnated to suffer pain and suffering until we paid off the debts we owed.

In the past, these practitioners who hid in the mountains understood to get rid of the karma created in this life and past lives, there must be two steps. The first step is to stop creating more karma by not doing bad things. The second step is to do things according to the will of God, do good deeds, and suffer to reduce the karma of life after life. However, these "immortal" practitioners who hid in the mountains, because of their different cultivation methods, chose to avoid accumulating more karma and use suffering to reduce karma. This mechanism of reducing karma is very slow, so this mechanism must ensure the practitioners can live longer than ordinary people to eliminate all their karma in the long process of suffering. Finally, they can get rid of the flesh which imprisons them and achieve true immortality. This is why they can live hundreds or even thousands of years, but this is indeed a painful, slow, and clumsy way to free the soul and achieve immortality.

I think those wealthy and famous people who want to live a life of luxury and glory would not be able to endure such suffering. They would not give up everything they have and live such an ascetic life to find eternal life. However, being rich and powerful does not mean you cannot successfully obtain eternal life. In fact, it is very simple for rich people to obtain eternal life. But simple does not mean easy. So how simple is it? You only need to do three things.

The first step: wipe out the knowledge and definitions of the human body, life, and the universe that others gave you in the name of science on the big whiteboard of your mind when you were born. Make that part of your whiteboard the same as when you were just born, that is, a blank.

Step two: Like hundreds of millions of Falun Gong practitioners, pick up *Zhuan Falun* or the free online version and read it at least twice. Then go to the free counseling center in your city and talk to the volunteers there. Ask them why they want to practice Falun Dafa.

Step 3: After you can accept what is said in the book and what these people say, and before you reach the realm of eternal life, continue to read *Zhuan Falun* and strictly put the theories taught in *Zhuan Falun* into practice in your life.

It is so simple to achieve immortality, but at the same time, it is not easy because the hardest step is the first step...

So, what kind of book is *Zhuan Falun*? It is somewhat like the wonderful book of Chinese Taoist practice - *Taiyi Jinhua Zongzhi,* which is now known to Westerners as *The Secret of the Golden Flower*, a book about how to free the human soul from the bondage of the flesh and become an immortal. However, the method introduced in *The Secret of the Golden Flower* is a specific and detailed description of the inner elixir that is circulated in the small circle of Chinese Taoist practice to escape from life and death and achieve eternal life. The German translation by Richard Wilhelm ignored the requirements of this method on the practitioner's mind, that is, moral character or *Xinxing*, resulting in the practitioner only being able to see some light and golden flowers or mandala flowers at a very low dimensional initial stage. *Zhuan Falun* is a prehistoric practice method which originated from the source of the universe and is used to save all sentient beings. It is an important ideological principle of Falun Gong or Falun Dafa. This book has a guiding role for any orthodox religion or practice method, including all the practices in Taoism (*The Secret of the Golden Flower),* the practices in Buddhism, and the practices in Christianity. Chapter Five of this book will explain why Christianity is a practice method that can enable people to gain eternal life.

Below I will introduce my understanding of Falun Gong, a unique practice that helps people delay aging and ultimately achieve immortality.

Falun Gong originated from the fact that its founder, Master Li Hongzhi, first made Qigong public with extremely strong effects of curing diseases and keeping fit at the Oriental Health Expo held in Beijing in 1992. Master Li Hongzhi became the Qigong master with the most awards that year. The following year, in 1993, at the invitation of the government, he participated in the Beijing Oriental Health Expo again and became a member of the organizing committee. At the same time, he won the highest award of the Expo, the “Fringe Science Progress Award”, the conference’s “Special Gold Award”, and the “Qigong Master Popular with the Masses”.

According to official reports from the CCP at the time, due to Falun Gong and its effectiveness in curing diseases and improving health, 97 million people practiced Falun Gong in just seven years from 1992 to 1997. Other unofficial reports even said at least 100 million people were practicing Falun Gong in China at the time. However, because Falun Dafa’s teachings were completely opposed to communist ideology, it was regarded as a national “security risk” by the CCP. Since then, the CCP has launched a brutal persecution of Falun Dafa practitioners for more than 30 years, including forced organ harvesting.

Falun Gong reveals the universal characteristics of “Truthfulness, Compassion, and Forbearance”, emphasizing the importance of morality and goodness, and having remarkable effects in curing diseases and improving health. It does not rely on advertising but is spread in society through “person-to-person and mouth-to-mouth”. Practitioners generally have strong bodies and improved moral character. As of 2024, Falun Gong has spread to more than 150 countries and regions, benefiting hundreds of millions of people physically and mentally.

Falun Dafa is a practice which cultivates both the body and mind. As the level of cultivation increases, the body’s cells will gradually be filled with high-energy substances, showing a regress to youth. Finally, the body will be completely replaced by high-energy substances, and you can achieve eternal youth. At present, there are hundreds of millions of Falun Dafa practitioners at home and abroad, including many elderly people. Through cultivation, many elderly people have seen their gray hair turn black, wrinkles decrease, skin become tender, and energy and physical strength increase. Many elderly women after menopause have also experienced phenomena such as menstruation and other rejuvenation.

Modern medicine believes the birth, development, growth, survival, and death of multicellular organisms are natural and regular processes and human aging is an unchangeable law. The imbalance poisoning theory of aging believes: during aging, the body has multiple system imbalances, and various factors cause genetic errors and mutations, causing genes to express incorrect information and products, which in turn interfere with the normal metabolic process and form a vicious cycle, thus leading to aging. Therefore, the phenomenon of rejuvenation which occurs in Falun Dafa cultivation goes beyond the theory of modern medicine. However, it is not difficult to explain it using the principles of Falun Dafa. Because the universe has different time and space levels, the universe's characteristics "Truthfulness, Compassion, and Forbearance" have different manifestations at different levels, deriving different principles. Aging is irreversible for ordinary people, but when practitioners transcend the level of ordinary people, they are no longer restricted by the principles of this ordinary space. At the same time, practitioners collect a large number of high-energy substances from other dimensions through cultivation and store them in cells, which naturally inhibits the effects of molecular-level aging factors and slows down the aging of the body. During the cultivation process, after the degenerated cells are eliminated, the normal cells of the body are gradually filled with the cultivation power, or high-energy substances, which can achieve rejuvenation and even eternal youth. In Chinese history, many practitioners have left behind flesh bodies which have not decayed for thousands of years. When practicing Tibetan lamas well passed away, their bodies turned into rainbows. Those who practiced less well would leave some nails and hair, or their bodies would shrink into small flesh balls. Many Communist Party members who went to Tibet for work have witnessed this. People have always been puzzled by this. In fact, the reason is, the flesh bodies of these practitioners have been completely or mostly filled with high-energy substances collected from other dimensions during their cultivation.

Zhengjian.org reported in 2011, a photo taken accidentally showed the left arm and right hand of a Falun Dafa practitioner in the photo were transparent to varying degrees. (Internet photo)

Mr. Li Hongzhi said in *Zhuan Falun* that practitioners will experience a "crystal-clear body", and he states:

> *What comes next is a transitional phase between that of the human realm and that which is beyond, where you will have what is referred to as a "crystal-clear body." At this point, your body will still be a mortal one, even if transformed, but it will have reached the highest form possible while practicing in this realm. It is referred to as crystal clear for having achieved the greatest degree of purity and being made entirely of high-energy matter. Your whole body will be transparent if seen with higher vision, and look similar to glass, as if nothing were there. This means that your body is now a holy one, for it consists of high-energy matter, unlike the body you began with. (Zhuan Falun, p355, Li Hongzhi, 1994)*

According to Zhengjian.net, on January 11, 2001, a photo of an Irish Falun Dafa practitioner was taken while he was practicing in a park. The photo clearly showed a white light shield above his head. In another old photo, the camera captured a spectrum of light that is invisible to the human eye. A group of Falun Gong practitioners were meditating with a huge red energy ball hovering over their heads. Master Li:

> *My spiritual bodies sit in a circle there, and above the practice site is a shield, with a massive Falun on it, and a gigantic, spiritual body watching over the site from above the shield. This isn't like most sites where regular people gather to do exercises, but a place for spiritual refinement. Many among us with extrasensory powers have seen that wherever we gather to practice Falun Dafa the spot is enveloped by a shield of red light, and all within is bathed in this light. (Zhuan Falun, p149, Li Hongzhi, 1994)*

So if you want to achieve immortality, you have two choices. The first is to be like Richard Wilhelm and give up everything you have in the ordinary world, including your family and wealth, and go to the famous mountains and rivers in China to find a Taoist master and start a long and arduous practice which can last for decades or even hundreds of years.

Second, after finishing this book, go to Falundafa.org and read *Zhuan Falun.* If you feel you can accept the truth about the human body, life, and the universe, go to the free Falun Gong Teaching Centers in your cities and talk to the volunteers to learn how they became Falun Gong practitioners. I guarantee you will learn a lot from their life experience which you are not taught in school. Then decide what you want to do, but please rest assured they will not call you to harass you like some religious groups to ask you to join them, and no one will ask you for a penny.

You have also read Chapter 3 and understand karma is a negative energy. According to the law of conservation of energy, energy cannot be destroyed. It can only be transformed into other forms such as disease, pain, or misfortune; or transferred to other people and let others bear it for you. Only higher-level lives can bear these diseases and karma for humans.

The priceless methods for healing and the way to eternal life are granted to you by the sheer benevolence of Master Li Hongzhi and his disciples. He and his disciples came to this world not for money, but to save people by explaining the truth about the universe.

III. Who Is He (The Founder of Falun Dafa)?

I think after reading this chapter of the book, you must have a huge question in your mind, wondering who Li Hongzhi, the founder of Falun Gong, is?

According to Wikipedia, Master Li Hongzhi participated in the Beijing Oriental Health Expo in December 1992 and was the most awarded Qigong master. In 1993, he participated in the Beijing Oriental Health Expo and was also a member of the organizing committee. He won the Expo's highest award, the "Fringe Science Progress Award", the conference's "Special Gold Award", and the "Qigong Master Popular with the Masses". In 1994, he was appointed as a member of the Beijing Health Expo Organizing Committee and won many awards and commendations at the event. As of December 31, 1994, at the invitation of official Qigong Science Research Associations in various places, he held a total of 56 Falun Gong face-to-face classes in various parts of China, each lasting about ten days. His teaching of Falun Gong in various parts of mainland China ended in 1994. In the early and mid-1990s, Falun Gong and Li Hongzhi won many awards, and the official media reported positively, and the Chinese government also supported it. During this period, the official media reported on how Falun Gong benefited people physically and mentally and showed Falun Gong practitioners receiving the "Healthy Citizen Award". People from all walks of life in China, including doctors, farmers, workers, soldiers, and CCP members, practiced Falun Gong, including professors from Tsinghua University and employees of the official media Xinhua News Agency and CCTV.

During this period, Master Li Hongzhi had a positive relationship with the Ministry of Public Security. In 1993, he provided treatment to 100 police officers who had been injured on duty, winning praise from the Ministry of Public Security. Master Li taught classes at the Beijing Public Security University in 1994.

In April 1993, the first book, *China Falun Gong*, was published by the official Military Arts and Literature Publishing House. The book expounded the basic cosmology, moral system and practice teaching of Falun Gong. In December 1994, *Zhuan Falun* was published and distributed by the China Radio and Television Publishing House under the State Council's Radio and Television Department. The publication ceremony of *Zhuan Falun* was held in the auditorium of the Ministry of Public Security in January 1995. The book was listed as one of the "Top Ten Best-Selling Books in Beijing" by the *Beijing Youth Daily* in 1996. However, Ian Johnson, a reporter for the *Wall Street Journal*, pointed out during the period when *Falun Gong*'s books were the most popular in China, Li Hongzhi never received any copyright fees because the publications were pirated.

On March 13, 1995, Master Li was invited to hold a lecture and study class at the Cultural Affairs Office of the Chinese Embassy in France and went to Paris, France to teach the exercises for the first time overseas. From April 14 to April 20, 1995, the second overseas Falun Gong study class was held in Gothenburg, Sweden. This was the last face-to-face class. He also held three lectures in Gothenburg, one in Stockholm, and one in Uddevalla, each lasting about half a day. The three overseas teachings ended in April 1995, and his teaching came to an end. In 1996, he visited the United States for the first time, and in the same year he was awarded the title of "Honorary Citizen" and "Goodwill Ambassador" by the Houston City Government of Texas, USA for "selfless public service for the benefit and welfare of mankind."

In 1998, he settled in New York, USA as a permanent resident and has lived there ever since.

A long time ago, people asked Master Li Hongzhi, who is he? I remember Master Li Hongzhi seemed to answer like this, "Don't spend time trying to figure out who I am, because humans will never be able to understand who I really am." Then he said, "I am Li Hongzhi!"

People will understand who Master Li Hongzhi really is after he returns to China. But I can tell you that, just like Jesus and Buddha, Master Li Hongzhi possesses six qualities and abilities that ordinary people do not possess.

These six extraordinary qualities and abilities include Jesus' divine power, unparalleled cosmic knowledge, charisma which can completely change human nature, special birth time and place, unprecedented cruel persecution encountered to save mankind and described by all well-known Eastern and Western prophecies.

Below, I will list six qualities of Master Li Hongzhi which are unmatched and worthy of our deep reflection.

First, Unparalleled Divine Power

As we all know, the Gospels all record Jesus can restore the sight of the blind, make the lame walk normally, and heal lepers. Jesus also performed other miracles, such as walking on water, turning water into red wine, etc. Moreover, Jesus can bring those who do what he says to the kingdom of heaven established by Jesus.

At the two Oriental Health Expos held in Beijing in 1992 and 1993, Master Li Hongzhi and disciples personally took on the karma of thousands of patients with serious and terminal illnesses and cured them miraculously and quickly in front of the public. He once again showed the miracle of Jesus healing patients on a large scale to the world.

In addition, some readers call *Zhuan Falun* a magical book from outer space with healing effects. I am a beneficiary of this book after reading it.

Because I had been suffering from Comorbid intermittent explosive disorder and post-traumatic stress disorder (Comorbid IED and PTSD) for a long time, my bad temper ruined my marriage and career. No matter what kind of help I got or how hard I tried, I could not save myself from the pain. When I was about to end my pain by ending my life, I read *Zhuan Falun* by chance and after reading the book, I exclaimed in amazement, "Oh my God! How can there be such a book in the world? The knowledge in the book by no means comes from the human brain!" I found the answers to all the problems in my life. I also found the answers to questions about the Bible, religions, science, the human body, the universe, and a lot more.

The Comorbid IED and PTSD which had plagued me for decades disappeared after I spent only 25 bucks to buy a book. I want to tell you, you can read *Zhuan Falun* for free on the public welfare website of Falun Dafa. There are many cases like mine on the public welfare website where diseases were cured after just reading *Zhuan Falun* once. For example, a patient with severe edema urinated continuously during three days of reading the book, and the edema was gone on the fourth morning. There was also a stage 4 cancer patient whose cancer disappeared after reading *Zhuan Falun* for only four days. People were surprised to find not only the author himself has unparalleled healing power, but also the book he wrote has healing power.

The most important thing is that in the 1990s, thousands of qigong masters in mainland China taught health-preserving qigong, but only the qigong taught by Master Li Hongzhi could free people from life and death, and practitioners could obtain various special abilities through practicing Falun Gong, such as non-eye image recognition, remote viewing, perspective, prediction, telepathy, levitation during meditation, great strength, and soul traveling through different dimensions to see the truth of different dimensions. In fact, the special abilities of Falun Dafa practitioners are more diverse than those of Ms. Sun Chulin.

What I want to say is Master Li Hongzhi's unparalleled ability to cure human diseases and develop human potential is not an ordinary human special function, but a manifestation of divine power.

Second, Unmatched Charisma

A Chinese scholar once said:

> *In this chaotic world, it is unprecedented in human history to be able to make 100 million people change their minds, cultivate their minds and do good deeds, and to make tens of millions of ordinary people put aside life and death and stick to the truth. Now that Master Li has been framed, millions of Falun Gong practitioners risked their lives to petition at Tiananmen Square and call for Master to be cleared of all charges. This shows that Master Li's charisma is unprecedented in history.*

To put it in very simple terms, what did Master Li Hongzhi give these people that made them willing to sacrifice their lives to protect the reputation of their master? I think no one is stupid. Master Li Hongzhi must have given them something more than life, so they felt it was worthwhile sacrificing their lives.

Moreover, the teachings of Falun Dafa cultivation are not like the communist cult of the CCP, which brainwashes people from a young age, does not allow people to have the opportunity to contact other ideas, does not allow people to contact other religions, and ultimately achieves brain control over people. Falun Dafa is completely different. Many of the practitioners who believe in Falun Dafa are adults, and a lot of them are middle-aged and elderly people. They have rich life experience and have accomplished big achievements in what they do. Many of them used to believe in other religions including science, but in the end, in the face of one inexplicable miracle after another, they chose to believe in the highest truth of the universe-truthfulness, compassion, and forbearance. Many of them used to be terminally ill patients who had been sentenced to death by doctors, but they entered Falun Dafa cultivation with the idea of giving it a try. After practicing Falun Gong, not only did 98% of the patients' diseases get cured, but also, all practitioners believe they have become better people, and they always think of others when they speak and do things. They never fight with others for their own interests, and it is even more impossible to hurt others. They are always conscientious and hardworking workers at their jobs.

Falun Dafa practitioners are the kindest people in the world. If they are hit by a car in a traffic accident, they will not ask the driver who caused the accident unintentionally for medical expenses, nor will they sue others. They will let the driver who caused the accident leave without hesitation. Think about it, how abnormal is this behavior of completely considering others in today's society? If the same thing happened to an ordinary person, the driver would be sued, possibly to bankruptcy. The state of Falun Dafa practitioners who only consider others and have no self comes from the words and deeds of their master Li Hongzhi. This powerful influence on humans can only come from saints and gods.

Master Li Hongzhi's charisma is not only reflected in ordinary and kind-hearted people of faith but also in the tools used by the CCP's dictatorship to suppress the people, namely the CCP's police and army.

In 1993, Master Li Hongzhi treated 100 police officers who had been injured on the job and won recognition from the Ministry of Public Security. The Ministry of Public Security, the country's police system, was very grateful for Master Li Hongzhi's contribution and requested that Master Li Hongzhi teach Falun Dafa in the police school. In 1994, Master Li Hongzhi opened a class at the Beijing Public Security University.

Master Li Hongzhi's first book, *China Falun Gong*, was published with the support of the inspired Chinese Communist Army. Many people in the Chinese Communist Army system practiced Falun Gong. So in April 1993, China Falun Gong was published by the Military Arts and Literature Publishing House, which was affiliated with the military. This book expounded on the basic cosmology, moral system and exercise teachings of Falun Gong.

Subsequently, in December 1994, *Zhuan Falun* was published and distributed by China Radio and Television Publishing House, which is affiliated with the Ministry of Radio, Film and Television of the State Council. Upon learning *Zhuan Falun* would be published in 1994, the Ministry of Public Security took the initiative to propose that the Ministry's auditorium be used free of charge as the venue for the publication ceremony as a token of gratitude.

Therefore, when *Zhuan Falun* was published, the Chinese police system, which was inspired by the charm of Master Li Hongzhi, also contributed to it. In 1996, the book was listed as one of the "Top Ten Best-selling Books in Beijing" by the *Beijing Youth Daily*.

However, Master Li Hongzhi did not make a fortune from publishing these two books as most people imagine. During the period when Falun Gong books were selling the most in China, Master Li Hongzhi never received any copyright fees because his publications were pirated.

When asked about this, Master Li Hongzhi explained he came to save people, not to make money. To this day, Master Li Hongzhi still lives a frugal life relying on his royalties. Even so, he still takes out a large part of his royalties to give pocket money to the students of Feitian University regularly.

A disciple who followed Master Li Hongzhi revealed, during his time in China, Master Li Hongzhi would buy cheaper instant noodles in Beijing and only eat these noodles on the train and in the hotel for each meal of the day so he could save travel expenses to lower the cost of his classes.

His disciples said they wanted to vomit whenever they saw instant noodles because they followed their master and ate instant noodles every day, every meal.

Once, Master Li Hongzhi saw what was on his disciples' minds, so he took them to a restaurant and ordered a table full of delicious food for them. Just as the disciples were wolfing down the food, they discovered their master had not eaten but was just sitting there peacefully, watching them eat with a smile, as happy as a father watching his hungry child eat.

The disciples knew the master must have eaten instant noodles in the hotel before inviting them to dinner so they didn't need to ask why he didn't eat the delicious food. He wanted to reduce expenses and the cost of classes for new students while letting the disciples eat as much delicious food as possible.

No wonder the disciples around him often helplessly "complain" Master Li Hongzhi is good to everyone but not himself. Master Li always teaches by example and always considers others in everything he does, even when dealing with his own disciples.

However, Master Li Hongzhi's unparalleled charisma caused dissatisfaction, jealousy, and panic among the then-powerful Jiang Zemin. Since the Jiang Zemin era, the Chinese Communist regime has been carrying out an unprecedented and brutal persecution of Falun Gong for 25 years, including the live organ harvesting of Falun Gong practitioners.

Third, Unprecedented and Brutal Persecution Continues

Over the past 25 years, tens of thousands of Falun Gong practitioners have been persecuted in China. The CCP's persecution of Falun Gong includes forced labor, torture, and even live organ harvesting.

On the afternoon of June 25, 2024, the U.S. House of Representatives unanimously passed the Falun Gong Protection Act (H.R. 4132) by voice vote. The bill requires the United States to stop the CCP's state-backed campaign of live organ harvesting from Falun Gong practitioners and other prisoners of conscience and requires the United States to impose sanctions on those who participate in and assist in live organ harvesting in China. Sanctions include freezing property in the United States, prohibiting entry into the United States, and a series of economic and prison penalties.

Fourth, His Special Birthplace and Era

Master Li Hongzhi was born in Communist China. Why did Master Li Hongzhi choose to reincarnate in CCP's China?

- Think about what kind of country China is today, ruled by the Chinese Communist Party?

- Do you think the war between Israel and Hamas which broke out in 2023 has nothing to do with the CCP?

- Do you think the war between Ukraine and Russia which broke out in 2022 has nothing to do with the CCP?

- Do you think the new coronavirus (CCP virus) which has killed hundreds of millions of people and has had a huge impact on the global economy since the end of 2019 and has continued to this day has nothing to do with the CCP?

- Do you think North Korea and Iran's nuclear weapons capabilities have nothing to do with the CCP?

- Do you think the possible outbreak of World War III in the future, that is, a nuclear war which may destroy all mankind, will have nothing to do with the CCP?

Although there are more than 190 countries in the world, only China and the United States can decide the fate of mankind. China, which is occupied by communists, not only did not take the responsibilities which a major country should assume and actively influence the world, but instead stood on the opposite side of humanity and became an enemy of the United States, the only country that can take on the responsibility of saving mankind. The purpose of launching decades of Unrestricted War against the United States is to defeat the United States and dominate the world.

The ultimate goal of the CCP was to change the existing world order established by the United States after World War II. Just like the last wish of Mao Zedong, the founder of the CCP, to plant red flags all over the world, it means turning all countries in the world into dictatorships like China, North Korea, Iran, Venezuela, and Syria. And enslave the people of the world like they enslave 1.4 billion Chinese.

I clearly remember when I was in elementary school in China, my teachers always taught us there are some friendly countries in the world with very good political systems, including the Soviet Union, North Korea, Iran, and Cuba. At the same time, there are always some evil empires in the world, such as the United States, Britain, and Japan. As the successors of communism, your lifelong goal is to import communism into those evil empires especially the United States, to realize communism and "save" mankind.

Therefore, the root of all evil in this world lies in the Chinese Communist Party.

The time when Master Li Hongzhi was born was exactly the days before the final judgment mentioned in the Bible because the world was full of cities like Sodom and Gomorrah.

Both the New Testament and the Old Testament of the Bible predict the Messiah will come at the last moment of mankind. In their religious legends, before the Messiah comes to the world, one of the signs is the restoration of Israel, and the generation after the restoration of Israel can see the Messiah.

After the end of World War II, the Israelis, after thousands of years of wandering, finally regained their country under the watchful eyes of the world. On May 13, 1948, the Jewish Congress Declaration was issued in Jerusalem, announcing the "restoration of the State of Israel." Although the mainstream Western society is Christian and Catholic, and Israel is Jewish, controlling Jerusalem, which is one of the conditions for God's return, has always been so important the mainstream Western society has been very firm in supporting Israel for more than half a century, completely abandoning the historical religious disputes on this point.

In the East, there is also another important signal of the coming of the future Buddha Maitreya (the Holy King of the Wheel of Time). The eighth volume of the Buddhist scripture *Huilin Yinyi* states:

> *The Udumbara flower is caused by auspicious and supernatural phenomena. It is a heavenly flower that does not exist in the world. When the Tathagata descends to the world or the Golden Wheel King appears in the world, this flower will appear due to the great merit and power.*

The Buddhist scripture- "The Infinite Life Sutra" also records that the Udumbara flower is a sign of good fortune. Lotus Sutra Volume 4, Part 1:

The Udumbara flower is a sign of good fortune. It appears once every three thousand years, and when it appears, the Golden Wheel King will appear.

Since 1992, the holy Udumbara flower has been discovered in Korea, Japan, Taiwan, Thailand, Hong Kong, Malaysia, Singapore, Australia, various states in the United States, Canada, Europe, and various provinces in mainland China. People can see its noble and holy style through online image searches. The Udumbara flower has no roots, no leaves, no water, and no soil; it can bloom on glass, steel, Buddha statues, leaves, cardboard boxes, and plastic. Some flowers can bloom for more than a year and still be full of vitality. No one has seen it since ancient times, and today's botanists are also dumbfounded. According to Buddhist records, 2011 is the year 3038, and the Udumbara flower has bloomed all over the world, which is truly a sign from heaven. Many Falun Dafa practitioners often see the Udumbara flower blooming on the glass of their windows, and some have seen it as many as four or five times.

Be careful not to confuse the lacewing eggs with the Udumbara flower. The hatching period of lacewing eggs is four to seven days, but the Udumbara flower can bloom for up to a year.

Lacewing eggs are round and plump, and some larvae eat their own outer shells after hatching. They will even eat other unhatched larvae causing damage, stumps, and blood stains in eggshells.

If you look closely, the shells of lacewing eggs which have hatched look like discarded garbage bags of the same capacity. Over time, they will gradually turn yellow and gray, eventually becoming dull and even falling off. The Udumbara flower not only has a special fragrance but also a lasting holy luster. Its structure is asymmetrical and unique.

Therefore, the Savior will only be reborn in China before the CCP destroys humanity, because he knows where the root of all evil in this world lies.

Fifth, Predicted by both West and East Prophecies

A savior will come to the world to save all living beings during the last days of the dharma. The Buddhist scriptures believe the future Buddha Maitreya will come to the world to save all living beings during the last days of the dharma, while the Bible believes when the last days come, the Messiah will come to the world to save all living beings. If the Buddhist scriptures and the Bible are credible, then there will be two saviors for mankind, unless the Maitreya in the Buddhist scriptures is the Messiah in the Bible.

1: Maitreya Buddha and Messiah Are the Same Person?

An important contribution of the late famous Chinese studies master, Buddhist scholar, and translator Ji Xianlin and his disciple Professor Qian Wenzhong was to discover the connection between Buddhism and Christianity, that is, "Maitreya Buddha, the future Buddha of Buddhism, and Messiah, the savior of Christianity, are the same person."

According to the research of Professor Qian Wenzhong of Fudan University in Shanghai, around 1000 BC, a belief in a future savior was prevalent in a vast area including West Asia, North Africa, Asia Minor, the Mesopotamian Basin, and Egypt. The Messiah in the Jesus religion is the most representative of this belief in the savior. This belief already existed in the Old Testament of the Bible. The Indian Maitreya belief has been confirmed in the academic community to be closely related to this worldwide belief in the savior and to influence each other. The Indian Maitreya belief is a component of the belief in the savior. In the simplest terms, the reason why Maitreya is the future Buddha, and the future savior has its roots in India, as well as in the wider world or the ancient world is that it is a part of the Messianic belief that was popular at the time.

Maitreya in Sanskrit is called Metteya in Pali. The English translation of the God awaited by the West is Messiah, which is translated from the Hebrew Masiah (sometimes written as Mashiach). The Greek translation is Christos, "the anointed one", from which "Christ" is derived. "Messiah" and "Christ" basically have the same meaning, and the author of the New Testament also equates him with the Jewish Messiah.

Maitreya and Messiah sound close to each other. In fact, Maitreya from Tocharian is the Messiah in Hebrew. It is the same word, but in the West, Messiah is pronounced while in the East, Maitreya is pronounced. It's been very common for a similar situation to occur in the history of human civilization.

2: Prophecy of French Prophet Nostradamus

In his book "*The Centuries*", a collection of prophecies, the Frenchman Nostradamus accurately predicted many major events and figures that have occurred around the world over the past few hundred years. Among them, the 29th chapter of the Second Age, "*The Centuries*", clearly predicted the savior who would save all living beings in the last days of mankind.

> *The Easterner will leave his seat,*
> *To pass the Apennine mountains to see Gaul:*
> *He will transpire the sky, the waters, and the snow,*
> *And everyone will be struck with his rod.*

On March 13, 1995, Mr. Li Hongzhi, the founder of Falun Gong, was invited to hold the first Falun Gong class in France. This class introduced Falun Gong to the West and marked the beginning of its globalization. Today, people of all ethnicities practice Falun Gong in more than 150 countries and regions around the world, and the principles of "Truthfulness, Compassion, and Forbearance" have spread throughout the world.

Nostradamus mentioned "great Hermes" in the third line of the 70th poem of the 10th century to describe the founder of Falun Gong's spreading of Dafa as a way of explaining and spreading the teachings of the gods and Buddhas of the universe to mankind. "Hermes" has a magic rod he uses to awaken the world, so the fourth line of this prophecy says, "Everyone will be moved by his magic rod," which predicts the spread of

Falun Dafa will move everyone in the world, and their attitude toward Dafa will determine their future.

It is well known that "Centuries" predicted many major historical events, all of which came true. What is less known is this great prophet Nostradamus has an unsolved mystery with Mr. Li Hongzhi.

3: Jewish Father and Son: Our God Has Asked Us to Invite You to France to Save Europe

From July 1 to 8, 1994, Master Li Hongzhi held a Falun Gong class in Dalian, Liaoning, China. During the class, a phone call came from Beijing saying some French people had come to China to seek an interview. Master Li said, "I know they are here to treat their children. Tell them to find another Qigong master!"

After a while, the phone rang again, saying they must meet and were ready to go. Master Li Hongzhi listened and said, "Come if you want to. If you have a chance, let's meet!" Soon, the people from France arrived in Dalian. Master Li Hongzhi met them in the hotel. The translator first introduced the identities of the two people, saying that they were father and son, Jewish, naturalized in France, and had a large group of industries.

The father spoke first, "We are Jews. We know that humanity is in great danger today and will face a doomsday disaster. Our God told us that only one Chinese person can save humanity, the French, and the Jews. We have investigated for a long time and we know that it is you–Mr. Li Hongzhi, the Chinese Qigong Master. So, we came to see you!"

He spoke very directly: "Our God asked us to invite you to go to France to save Europe!" He said with a pleading tone and a pious expression: "We will take full responsibility for all the procedures and expenses for your trip to France." After listening to him, Mr. Li Hongzhi said calmly, "Tell me about anything else!"

4: Master Li Hongzhi Remotely Diagnosed a Sick Child in France, Paving His Way to the West

The young Frenchman said he had an 8-year-old son who was born with a dull brain. He couldn't speak, couldn't move freely, and drooled. He had sought medical treatment in many countries but to no avail, so he begged Master Li for treatment.

After a brief conversation, Master Li drew a human figure in front of the bed with his hands and stared at it for a while. Then he pulled the head of the human figure outwards with his hands, as if pulling out a diseased thread, and pulled it several times in a row. Then he closed his hands, picked up the teacup on the table, opened the lid, put the thing in his hands into it, and then closed the lid. Then he patted the human figure from head to toe three times. Then, looking at it intently, said: "I think it's healed. There should be a reaction there during this period of time. You can call and ask."

However, Master Li Hongzhi was staying in an ordinary small hotel (Wuhan Hotel), the conditions were not very good, and international long-distance calls were not available at that time. Before the two French people left the hotel, Master Li said, "Come, I will let you experience Falun. Stretch out your hands." Then, he drew a circle in the air with his right hand in their direction. After about ten minutes, Mr. Li Hongzhi asked the father and son how they felt.

The father said, "A force surged from my palms into my body. It was very strong and hot."

His son said, "My whole body was shaking. The force was very strong. My hands were the strongest. I couldn't let go even after 15 minutes."

Three hours later, the younger French called back and said in a choked voice, "I just called France, and my wife said, "At around 4 o'clock, I was suddenly awakened by a powerful force, and the room was very bright. I thought maybe my husband found the Chinese qigong master who was working on my son remotely. I immediately ran to the my son's room, and as soon as I entered the room, I saw the my boy sitting up on the bed. When he saw me come in, he asked, "Mom, what's wrong with me? "His pronunciation was clear and accurate, which had never happened before. My wife hugged my son and burst into tears excitedly."

The translator said, "They want to meet Master Li in the evening and are anxious to return to France tomorrow." That night, Master Li Hongzhi had dinner with the two from France. These two French Jews were very grateful and excited that Master Li cured their family member's illness and they invite Mr. Li Hongzhi to France to teach the Falun Gong.

Finally, on March 12, 1995, Falun Gong founder Master Li Hongzhi was invited to hold a Falun Gong lecture in the auditorium of the Cultural Affairs Department of the Chinese Embassy in Paris, France. The Chinese Ambassador to France Cai Fangbai and his wife were also in attendance.

Since then, Falun Dafa had officially started to spread overseas. This French Jewish father and son opened the prelude to Falun Dafa going global.

5: Nostradamus Was the Jewish God Who Sent the French Father and Son to China

Let's go back to the night when Master Li Hongzhi had dinner with two Jews and their translators.

During the dinner, a Falun Gong practitioner from Dalian (who was also a witness and recorder of the entire incident) took a few photos with his camera. However, after the photos were developed, the practitioner saw there were many unclear images on the wall in the photos. Master Li explained to him:

> *Many gods have come, including their Jewish God. Do you know who the Jewish God is? It is Nostradamus, whom you know. Ninety-nine percent of Nostradamus's prophecies from more than four hundred years ago have come true. He came to warn people that there will be a great disaster, so people will believe his final prophecy of the destruction of mankind. This disaster is set to happen, but when I come to teach the Fa, I cannot follow the original arrangement. I will change it. The great disaster will no longer exist, but the bad people will be cleared out. Too many people have died, which is quite terrifying.*

The entire process of the "remote diagnosis of the French sick child" incident described above can be found in the article "Days with Master (V)" published on Minghui.org on November 6, 2015. The author is a Falun Gong practitioner in Dalian who attended the first and second Fa-teaching classes held by Master Li in Dalian on March 27 and July 1, 1994.

6: Casey: China Will Become the Cradle of New Spiritual Beliefs

Edgar Cayce (1877-1945) of the United States was known as the "sleeping prophet" because he was in a state like "sleeping" every time he interpreted cause and effect and predicted the future. He was recognized as an outstanding prophet in the 20th century. Cayce's predictions about China during his lifetime shocked Westerners the most. Especially this major event which has been happening from the late 20th century to the 21st century. He said:

> *One day, China will become the cradle of the Messianic faith (Christianity), and people will strive to practice this faith in their lives. Yes, this is very far away for people, but in the heart of God, it is only a day's work–tomorrow, China will wake up!*

7: Cayce's Predictions Further Confirmed

Mark A. Thurston, Ph.D. is a psychologist, educator, and author of more than a dozen books about spirituality, integrative health, and explorations of the mind. Thurston is an expert on Cayce's predictions and mentioned nine of Cayce's predictions which have yet to be verified in his book Cayce's Prediction for the 21st Century. According to the author's observation, at least the following are related to the above-mentioned Chinese predictions:

*The Savior will come again.

*The truth of the universe's "One Mind" (the Creator) will become the basis for all human activities (including science). The new belief is the light of the universe.

*A new approach to medicine will emerge–based on the flow and transformation of spiritual and physical energy systems.

Falun Dafa practitioners work hard to cultivate in their daily lives, improve their moral standards, and purify their bodies and minds. According to empirical research, Falun Gong has a 98% chance of curing illness and improving health for practitioners and is also effective in treating cancer (Five Surveys in 23 Years: Confirming the "Supernormal" of Falun Gong). This reality has confirmed the mystery pointed out by Mark Thurston:

> *A new medical method will emerge–based on the flow and transformation of spiritual and physical energy systems.*

The *Bible, John* (1:10-11):

> *He was in the world, and the world was made through him, and the world knew him not. He came to his own, and his own did not receive him.*

The Bible also says when the Savior comes to the world to save people, his light will instantly illuminate the world, like lightning that flashes across the sky from one end to the other, but the world does not think of him as the Savior, and he must first suffer a lot and be slandered by people who do not know the truth.

The prophecy in "*The Centuries*" also uniquely marked the year and month of this major event: "In July 1999, in order to revive King Angora, the King of Terror will fall from the sky..."

July 1999 was indeed a special day. Jiang Zemin, who was labeled as the Antichrist by The Century, launched a comprehensive crackdown on Falun Dafa practitioners. For a time, there were bloody storms, slander rumors, and tribulations in the Central Plains, confusing the world and making people follow suit. This final battle between good and evil, Armageddon, described in the Bible Revelation as the lamb and the beast, has been nearly 25 years. While countless

Falun Dafa practitioners have endured nameless tribulations with perseverance, they have never given up their mission to tell the truth and save sentient beings. The CCP openly opposed the universal Buddha Dharma of "Truthfulness, Compassion, and Forbearance" and has been deeply mired in this battle between good and evil. The prophecy of "Ninety-nine years will make a big mistake" in the Tang Dynasty's "Tui Bei Tu" has come true.

8: Jesus also Left Clues for Mankind to Find the Messiah

On May 13, 1951, the eighth day of the fourth lunar month, Master Li Hongzhi was born in Gongzhuling City, Jilin Province, China, in the Year of the Rabbit. This day is exactly the "Buddha's Birthday" in Buddhism; it is also the 50th day of Easter 1951, which is almost the "Pentecost" in Christianity. At this point, people with insights may have a new understanding of why the important festival of Western religions, "Easter", is called "the one coming from the East", and why people have always retained the ancient custom of using the Easter bunny (the saint is born in the Year of the Rabbit) and Easter eggs (related to the Chinese golden rooster) for commemorative activities.

The Christmas tree every family puts up at Christmas is actually a small pine tree. If you regard the Chinese character "李", the last name of the Hongzhi Li, as a compound Lego building block, "李" can be split into two small Lego bricks or Chinese characters "木"-wood and "子" -son.

"木"-wood represents pine trees because, in the five elements of Chinese culture, pine trees belong to wood and are composed of wood.

"子" means children and offspring, and here it refers to a young pine tree, meaning that the saint's surname is hidden in it.

If you stack the above two Lego bricks of Chinese characters together, you will get a combined Lego block "木" + "子" = "李", which is the surname of the saint Master Hongzhi Li-李 in Chinese characters.

The most important thing is that the tree representing the Chinese character "木" is not any other tree, but the pine tree representing Christmas. The implicit meaning here is that only the Christmas tree is worthy of representing his surname, so he is a saint.

Liu Bowen, known as the Chinese Nostradamus, also predicted that the "future Buddha" Maitreya would be reincarnated in China in the Year of the Rabbit and would be born in the place where the eyes of the "Golden Rooster" in China are located (referring to Jilin Province), with the surname "Li," The hints in this prediction make people feel the mysterious will of God.

Some people may ask, if Master Li Hongzhi is the savior, why don't the Bible and Buddhist scriptures mention his name directly?

Because the laws of the divine world are much stricter than the laws of our human world. Gods cannot casually reveal their secrets. If they do, they may fall and be punished in the prison of Earth-like humans. People can only learn from puzzles and use their independent analytical thinking to make judgments.

Sixth, His knowledge of the Human Body, Life, and the Universe Is Unmatched.

After reading his major work *Zhuan Falun* and the articles he published in previous Fa lectures, many readers felt they had found the answers to the questions about religion, the human body, life, and the universe which had puzzled them for a long time.

Since I am not qualified to talk about this knowledge, I have specially selected some of the truths Master Li revealed in previous lectures in Chapter 6, including the structure of the universe, science, religious beliefs, higher-dimensional lives (gods, Buddhas, and Taoists), human destiny, extraterrestrial life, artificial intelligence, and other issues the public is more concerned about.

If you want to know more about the truth, you can go to Falun Dafa's official website Falundafa.org, and search for questions which interest you in the transcripts of Master Li's lectures at previous Fa conferences.

CHAPTER 5:

The Answers to Your Life

I have selected and reprinted some topics of interest to humans today from Master Li Hongzhi's past lectures and teachings, such as the origin of mankind, the structure of the universe, science, religious beliefs, high-dimensional life (gods, Buddhas, and Taoists), human destiny, extraterrestrial life, artificial intelligence, etc.

If you want to learn more about the truth of the human body, life, and cosmos, please visit falundafa.org.

I. How Humankind Came to Be (By Li Hongzhi)

"I would like to first pass along my greetings to everyone on the advent of the Chinese New Year.

New Year's would normally be a time for sharing a few pleasant remarks about the occasion. But I am seeing imminent danger approaching humanity, and have been called upon by divine beings to pass along, for this reason, several things to everyone in this world. Each of what I am about to disclose is a higher, closely guarded secret, and these are being shared to provide a true picture of affairs, and to give people another chance at salvation.

First is the question of how humankind came to be. From the dawn of its creation to its final days, the universe has gone through an exceptionally long passage of time consisting of four stages: Formation, Stasis, Degeneration, and Destruction. When the final point of the Destruction stage is reached, the complete obliteration of everything in the greater cosmic body—which includes the universe in which we exist—takes place instantaneously, and all living things perish!

When a person dies it is just a matter of his physical body declining and breaking down, while his true soul (which is who he really is, and which does not die with the passing of his physical body) will continue on in a next life, being reborn. So just as the universe goes through Formation, Stasis, Degeneration, and Destruction, so too do human beings go through birth, aging, illness, and death. These are laws of the universe, to which even higher beings are subject, only the time span is longer, with the process being more drawn out in proportion to how great the beings are. Life and death are not painful for them, and they remain cognizant throughout these processes—to them it is as if but changing outfits. Put differently, normally lives do not really die. But when the universe and the cosmos disintegrate at the final stage of the Formation, Stasis, Degeneration, Destruction process, however, lives will not be reborn, and there will be no more existence of life or matter, with all turning to dust and there being only emptiness. Currently the human world is experiencing the last period in the Destruction phase of the Formation, Stasis, Degeneration, Destruction progression. Everything has changed for the worse in these end times, as fated, and destruction is thus imminent. And it is for this reason that the world is so troubled. Good thoughts are rare, people's minds have become twisted, debauchery and drug abuse are rampant, and people subscribe to atheism.

These are inevitable in the last stage of the cosmos, and bespeak of the time at which we have arrived!

The Creator cherishes all of the heavenly beings that exist as well as all of the lives that are good and kind, and all of the glorious creations in the cosmos. So at the beginning of the Degeneration stage He led a number of divine beings to the outermost plane of the cosmic body (known generally as "that which is outside the Divine Realm"), a place where there are no divine beings, and created Earth. But earth hadn't the capacity to exist independently; it needed for there to be a corresponding cosmic structure with which it could form a circulatory system involving life and matter. For this reason, the Creator made a larger expanse outside of the Earth, which higher beings refer to as the "Three Realms." Before the final time of salvation would arrive, no higher beings, however great, were to be allowed to enter or exit this expanse without the Creator's permission. The expanse of the Three Realms comprises three major realms: the Realm of Desire (yu), which is made up of the lives on this earth, including humankind; a second realm, the Realm of Likings (se), which is above it; and a third realm, further above, known as the Realm Without Likings (wu se). Each successive realm is higher and more glorious than that below it, though none can compare to the Divine Realm or the many heavenly kingdoms still higher. The "heaven" that people normally refer to is in fact within either the Realm of Likings or the Realm Without Likings, within the Three Realms. Each of the Three Realms has ten planes within it, making for a total of thirty-three planes in all, if you include the Three Realms themselves. Human beings reside in the Realm of Desire, and this is the lowest of all planes, with the worst environment. Life is painful and short here, but more dreadful still is the fact that in the human world few of the things people take to be truths are actually valid. What human beings hold to be true is on the whole considered the opposite in the larger universe (but an exception is the higher truths that holy beings have taught to man). For example, the divine does not consider it right for whomever is victorious in battle to become ruler, for territory to be seized by military force, or for the powerful to be seen as heroes—since killing and forcefully taking from others are involved. That is not the way of the universe, nor how higher beings go about things. Yet in the human world these are inevitable, and accepted. Those are the ways of the human world, but they are contrary to the ways of the universe. Thus, if a person wishes to return to heaven he must follow true, higher laws and work on himself. Some people are content when they are doing a bit better in life than others. But such people are only comparing themselves with other human beings within this human realm, when everyone here is in fact living in what is considered to be the trash bin of the universe. The Three Realms were established at the outermost perimeter of the cosmic body, and everything here is made up of the lowest, crudest, and filthiest of particles—molecules, atoms, and the like. In the eyes of higher beings this is where the trash of the universe is cast away. They thus regard this plane of molecules as dust or "clay," and see it as the lowliest of places. This is the origin of the belief held in some religions that man was made by the divine out of clay. Man was indeed formed out of matter at the molecular plane.

When divine beings made man, they did so at the Creator's behest, and He instructed them to each make human beings in their own unique image. For this reason there are the White, Asian, Black, and other races. While their outward appearances differ, the souls within them were given by the Creator. And that is why they have common values. The purpose that the Creator had in directing divine beings to make man, was to make use of man in the final times when He would offer all lives of the greater universe—including heavenly beings—salvation.

But why would the Creator have divine beings create humans in such a lowly and inferior place? It was because, with this being the universe's lowest plane, it is the most grueling of places, and only when things are trying and painful may a person elevate himself through spiritual practice and shed his or her karma. When a person, amidst painful experiences, still manages to keep kind thoughts, have gratitude, and be a good person, he or she is growing through it. Salvation is a process of ascending from low to high, and so one has to start from the bottom. Life is trying for anyone living here. There are the tensions between people when they are trying to make out better, there is the terrible state of the natural environment, and there is the fact that just getting by in life takes a great deal of thought and effort, to name just a few examples. All of these circumstances provide people with opportunities to develop themselves and lessen their karma. It is certain that going through hardships can help people to atone for their sins and karma. And anyone who manages to stay good-natured amidst painful situations and interpersonal troubles is going to build up merit and virtue and, as a result, will achieve the elevation of his or her soul.

With the arrival of modern times, the Creator had intended to utilize mainly the human body to save the many lives of the universe. And so the souls originally in the majority of human bodies here were replaced by those of higher beings, who incarnated in them. With a human body they could reduce their karma and sins by enduring hardship. And in this place that is devoid of truth they could, by holding fast to the higher truths taught by God or holy beings, and persevering in goodness and kindness, achieve the elevation of their souls. The end times are now upon us, and the Heavenly Gate that leads out of the Three Realms has been opened. The Creator has been selecting for deliverance those who have done as I describe.

Everything in the universe had become impure during the Formation, Stasis, and Degeneration phases, and inferior to when creation was begun. And this is why things have been heading for Destruction. In other words, everything in the greater universe has gone bad, the lives of creation are no longer as good as they were in the very beginning, are no longer pure, and all of them have accrued karma and sins. And this accounts for Destruction coming about. This kind of sin is what has been referred to in religious contexts as original sin. So that the universe could be saved, the Creator directed a multitude of higher beings and divine sovereigns to descend to the earth and assume human form in this setting, where they would suffer, elevate, atone for their sins, and forge themselves anew—re-ascending to heaven as a result. (The Creator has been re-making the universe at the same time as saving humankind.) The new universe is perfectly pure and simply glorious. If, in a trying setting like this, a person can still keep his thoughts virtuous; if he can hold his ground against the onslaught of modern values and views, and stick to traditional ones; and if he still believes in the divine in the face of assaults from the atheist and evolutionary camps, then that person will fulfill his purpose: to gain salvation and return to heaven. All of the madness now unfolding in the world was planned as such, for the final phase, by divine beings. Their goal was to test the lives here and see whether they were worthy of salvation, and give them a chance to, in the process, work off their sins and karma while going through difficult things. And all of this was done so that people could be saved and gain deliverance back to heaven.

This is to say that the purpose of people's lives on this earth isn't to accomplish something in the world. All of the intense efforts and attempts people make in life, and their drive to get what they want, which can even involve resorting to unscrupulous means, only make people immoral in the end. The reason people came to this world and became human was to atone for their sins and karma, and to make significant spiritual progress. People came to this world to gain salvation. They came and assumed human form to await the Creator and His salvation back to their heavenly kingdom. And while they waited, they built up merit over their many past lives; and that was the purpose of people's rebirths. The troubled nature of this world is meant to make something great of these lives. Of course, there are some people who, when seeking divine help in times of duress, haven't been satisfied with the outcome, and started to loathe God—even turning against Him as a result. Some have even turned to the demonic, dark side, and committed still further sins and made yet more karma. Those whom this applies to had best quickly come around and beg God for forgiveness, if they are still to have a chance of reaching safety. Everything that happens in one's life—whether it seems warranted or not—is, in reality, the karmic consequence of what one did in one's past lives, for better or for worse. The amount of blessings and virtue that one built up in one's past lives determines what fortune is in store in this lifetime, or perhaps the next. If one lives a blessed and virtuous life now, perhaps it will translate in one's next lifetime into a high position and salary, or it might translate into different kinds of wealth and fortune. And this would also include whether one has a happy family, or even how one's children turn out, and so on. This is the fundamental reason why some people are wealthy and others poor, why some hold positions of high rank while others are destitute and homeless. It's nothing like the diabolical nonsense that sinister communism spouts about equality between rich and poor. The universe is fair. Those who do good are blessed for it, while those who do wrong will face payback—if not in this life, then in the next. For this is an immutable law of the universe! Heaven, Earth, the Divine, and the Creator alike are compassionate toward all lives. Heaven and Earth, just as with man and the divine, were made by the Creator, and it is never the case that He plays favorites with some lives and shortchanges others. The reason some people lead happy lives and others do not all comes down to rewards and retributions for past deeds.

When you see people winning or losing in life, it appears to come about in a normal way from things in this world. But it is ultimately the karmic consequences of those people's past doings. Whether people have something or not, or are winning or losing in life, is going to play out in ways that accord with this world. So no matter whether you are rich or poor in life, you should be sure to do good, refrain from doing wrong, stay good and kind, be spiritual and devout, and be happy to help others. And by doing so you will build up blessings and virtue, and reap their rewards in the next life. In the past, the older generation in China would often talk about things like not lamenting your lot in life when things are hard, and about earning a better next life by gaining virtue through good deeds. And the point was that it's useless praying to God for help if you didn't do good things in your former life and earn blessings. The universe has its laws, and even higher beings must obey them. Even they will be punished if they do things they shouldn't. So things are not as simple as people take them to be. Should people expect higher beings to give them whatever they pray for? The prerequisite is that one has to have built up the blessings and virtue for it over past lifetimes. And so the things that come to you are on account of the blessings and virtue you have! This is what the laws of the universe dictate. But speaking on a fundamental level, getting what you want is not the ultimate goal of accumulating blessings and virtue. The real purpose of building those up is to pave the way for you back to heaven. And that is what's most crucial, not the brief round of happiness that they can bring you in this lifetime!" (How Humankind Came To Be, Li Hongzhi, January 20, 2023)

II. The Structure of the Universe (By Li Hongzhi)

"While on this topic, let me take the opportunity to elaborate, from another angle, on the topic of dimensions. I just said that human beings come from different dimensions and levels. I will talk about the structure of the cosmos in passing. As I have said, our Earth is located at just about the center of the cosmos. An extremely miniscule number of planets like Earth exist elsewhere, but only our planet is at the central position. Occupying a central position has its significance, but what that means in a cosmic sense hasn't been discussed. For example, when ordinary people

acquire knowledge in human society, they regard the center as the highest or the best. I can tell you that in the context of the cosmos, however, Earth, being in the center, is in the worst position. Why? Because the cosmos is spherical and various kinds of lives in the cosmos will drop. Where do they go when they fall? Since the cosmos is spherical, while the left is the top, isn't the bottom also the top? The right is also the top (making hand gestures), as is the rear, and the front, so the bad things in the cosmos will fall downward. And where do they go when they fall? They fall to the center, don't they? Yet the cosmos is extremely complex, and there are still other concepts related to the central position.

There are so many levels in the cosmos. For example, roughly less than 3 billion galaxies like the Milky Way comprise the expanse of one universe. We usually call this a small universe. Then there are about three thousand such small universes beyond this one, and the distance among them is quite far. There isn't such a concept among ordinary man. Even when common gods conceptualize it, that distance is still considerably long and beyond the range of vision. Yet there is a shell outside of these three thousand universes that constitute a second-layer universe. Going further in this manner, beyond a second-layer universe, three thousand universes of that size make up a third-layer universe. The concept these numbers describe is enormous. If we use a grain of rice to represent one universe, then when we multiply it out, things get considerably big. Then three thousand grains of rice get to be as large as a desk, or even larger. Then three thousand desks, in turn, couldn't be contained by this auditorium. So the size of the multiples is considerably vast. In other words, its volume and scope are extremely immense.

Yet what I have told you about this cosmos is merely a simplified conception. A universe such as this one is still seen as one particle. In reality, it is indeed but a particle, and it is one type of small particle in the cosmos. Also, in the overall composition of the cosmic body, within each particle are its cosmic body's different forms of existence; there are also different layers of heavens within that cosmic body. Within each particle of a cosmic body there are different layers that, to the beings inside them, are different heavens or different layers of heavens. Within these different layers of heavens are the innumerable paradises of different gods. Last time when I came to the United States and taught the Fa in San Francisco, I told you a concept that was enormous. How many layers did I mention? At first I spoke of eighty-one layers, and later on I spoke of over one thousand layers. That expanse which I talked about was quite vast. But I can tell you, the expanse I talked about, which you consider inconceivably vast, is in fact still extremely small. If you were to be at a distance from it, or rise above it, and look back at it, it too would look like a speck of dust in the cosmos. The cosmic body is that vast.

Yet I am telling you that the gigantic cosmic body that develops ("develop" isn't the right word, but I can only use human language since there is no other language available) and expands with the Earth as its base—the connection there can only be described as "development"—is merely one system. Such systems are innumerable, and cannot be calculated with numbers known to man. There are different lives in those cosmic bodies, which are separated by great distances. Think about it. This cosmos is so immense that its size cannot be described with human language. So it follows that the differences among lives are substantial.

Humankind is always proud of the civilization it has developed so far, such as Manhattan having the highest and most numerous skyscrapers in the world. Humans are complacent with their present scientific achievements, which they consider amazing. Looking back, they think that the ancients were worse off than modern people; the ancients had to ride horse-drawn carriages, while today's people travel by car, train, and airplane. People think that their developments are quite glorious. But I will tell you this. You came from different cosmic bodies and worlds, and everything that has been developed in human society is the result of traces, or what's called "messages," from different cosmic bodies and societies that have remained in the human mind and enabled you to re-create such things among ordinary humans, using these coarsest materials of the human dimension. In other words, these things already existed long ago in different dimensions of the cosmos.(Teachings at the First Conference in North America, Hongzhi Li, March 29–30, 1998 ~ New York)

III. Contemporary Science (By Li Hongzhi)

"Even with the Dharma that Buddha Shakyamuni imparted, human beings have been able to sense the vastness of Buddha Fa. But the Fa is simply enormous—truly so. What Buddha Shakyamuni taught was, given his level, Tathagata Fa. And even that seemed very high to ordinary people. In the past, only a Tathagata could be called a Buddha. Neither a Bodhisattva nor an Arhat was a Buddha; neither had reached the realm of a Buddha. So it used to be that whenever someone said the term "Buddha," it was referring to a Tathagata. But Tathagata is not the highest status in the universe; rather, it is the lowest level of Buddha. And among the Buddhas, only those at the lowest level look

after ordinary human affairs. Buddhas of too high a level cannot do so, for the moment they opened their mouths, changes on Earth would occur. How could that be allowed? They can't get involved in the least. It is thus Tathagatas alone who look after human things. Tathagatas don't personally handle things, however. It is Bodhisattvas who do things to save people. So even just attaining the level of Tathagata is extraordinary.

At the microscopic level [a Tathagata] can perceive three thousand chiliocosms in a single grain of sand. In Buddha Shakyamuni's teachings there is the theory of three thousand chiliocosms. He said that in the Milky Way there were three thousand planets resembling that of man, containing lives identical to human beings. There are in fact more than three thousand. He also said that there are three thousand chiliocosms in a grain of sand. That is, in but a single grain of sand there exist three thousand worlds like this that man inhabits, something unfathomable. When the celestial eye is opened, one can see things as if they are magnified, and see tiny things such as those at the microcosmic level. But, just imagine this. There are three thousand chiliocosms in a single grain of sand. Then, the three thousand chiliocosms in that grain of sand have seas and rivers in them, don't they? Then, are there another three thousand chiliocosms in the sand of those rivers? So, even Buddha Shakyamuni had not seen the origin of matter. He thus said, "Its smallness is such that there is no interior," meaning that it is so small that the origin of matter is imperceptible.

Today physics can merely figure out that a molecule is made up of atoms, and that an atom is made up of a nucleus and electrons. Further into the microcosm there are quarks and neutrinos. And that's as far as it can go, that is all. Then what is, moving further downward, the origin of matter? Not even microscopes can detect quarks and neutrinos. Their existence is known only through the help of other instruments. As for answering what lies further down, man's technology falls woefully short. As to large things, man believes that planets are the largest objects. But that's not how Buddha Shakyamuni saw it. What Buddha Shakyamuni saw was in fact very sizable. He discovered that planets were not the largest. Peering further upwards, he couldn't see the zenith, however. In the end he thus said, "Its largeness is such that there is no exterior." So his statement that, "Its largeness is such that there is no exterior; its smallness is such that there is no interior," meant that it is so large there is no boundary to it, and so tiny that its bottom cannot be seen.

Living beings are complex, and the structure of the universe is extremely complex. The knowledge possessed by man today has reached its peak. Having peaked, it has become something that limits the development of man's science. For example, many accomplished scientists have set forth axioms in the fields of physics and chemistry. Within the confines of those axioms, what they say indeed holds true. And it's fine to build upon them. However, truth is a many-tiered thing. When you surpass their axioms, you will find that the axioms have in fact served to limit people.

Things are no different for the science of recent times. Some people set forth a definition of science, and so something is considered "science" only if it conforms to that definition. And when you stay within its boundaries, everyone thinks that is science. When you go beyond its definition, you discover that it has served to limit mankind's advancement. Nothing that is intangible or invisible is allowed for there, so the limitations it imposes are significant. The Buddhas, Daoist deities, and Gods we speak of exist in other dimensions that man cannot touch or see. Then if those beings were to be discovered using the methods of today's science, wouldn't that make them scientifically proven? It would! But the West has set forth a definition of science, and anything that modern science is unable to explain gets categorized as theology or religion, without exception. It dares not acknowledge such things.

Western science has gone to an extreme. The Buddhist school holds that everything goes through the stages of formation, stasis, and degeneration. Formation means taking shape, while stasis means remaining in a certain phase. The science from Europe, confined by the framework it established, now finds itself unable to advance further. Were it to keep probing downward, what might be discovered would be something beyond the boundaries of its science. So it categorically lumps these things together as religion or theology. But if someone discovers things that don't exist in the present body of scientific knowledge, and does so by way of man's modern science and technology, or discovers things that are intangible and invisible and studies them with the scientific method of our day, isn't that scientific? The problem is that the definition of science has long been set in stone, and anything beyond its boundaries is categorically denied. No room is thus left for further progress.

And there are some scientists, figures who are considered "accomplished" in certain fields, who have set forth a great many axioms. These scientists, such as Newton and Einstein, were very accomplished by the standards of ordinary people and could perceive far more than the average person. And the axioms they set forth, as with their scientific legacy, stands as a wealth of valuable knowledge. But that said, any research that takes place or understanding that's arrived at via working within the boundaries of their thought is bound to observe certain patterns. If those who come after them work completely within the theoretical frameworks of these scientists, posterity will never surpass them nor experience new breakthroughs.

When someone's discovery or invention surpasses the confines of previously held axioms, it will be realized that the axioms set forth had been restricting people. That's because there are higher forms of knowledge, and higher truths, to be found at higher levels. A good example is our knowledge of matter. It used to be that the smallest particle of matter known to man was the atomic nucleus. That's no longer the case, however, for now there are quarks and then neutrinos. The point is that human beings have continually learned more about such things. But a new axiom itself will, upon the discovery of something else, serve as yet another restriction. Such is the case. The fact is that such axioms usually serve to limit people.

Einstein was no ordinary person. He found what religion, and even theology, taught to be true. Man's understanding of the physical world is limited to the knowledge of human beings, much like the scientific axioms that have been set forth. Were people's research to truly probe deeper, and their endeavors to progress further, they would find what religion has taught to be true. The lives that exist on a plane one level higher than man thus represent a science and technology one level higher, and their understanding of the

world via the science and technology at their command surpasses that of ordinary human beings. That is why Einstein, upon having reached the pinnacle of human science and technology and then probing deeper in his work, found what religion taught to be fully real. In recent times many scientists and philosophers have ultimately turned to religion—and these are persons of quite some accomplishment. By contrast, those who are currently paralyzed by the limitations set forth by contemporary science and their blind faith in it categorically brand those things [beyond the reach of science] "pseudoscience."
(The Confines of Modern Science and the Breadth and Profundity of Buddha Fa, Li Hongzhi, First published 1996 Beijing, China)

IV. Western Orthodoxy (By Li Hongzhi)

"Religion is a name given by humans. When Jesus was alive, he never said that this is Christianity. When Sakyamuni Buddha was alive, he definitely never said that he was teaching Buddhism..It is humans who regard it as a part of ordinary human society and give it a name of "religion" and involve it in political affairs, and then call it a religion. It has lost the conditions that can truly enable people to cultivate and ascend, achieve Consummation and go to heaven." (Teachings at the Conference in Europe, Li Hongzhi, May 30–31, 1998 ~ Frankfurt, Germany)

"Many people have religious beliefs. But if you believe in today's religions, for one thing, gods no longer look after things, so it's hard for you to raise your level. Another thing is that a lot of people can't comprehend the true meaning of what was said by Jesus, Yahweh, or Saint Mary in their day. Some people hold that a person only needs

to have faith in order to go to Heaven. Having faith is only a foundation. Many people's understanding stops at the superficial meaning of the words. They say, "As long as I have faith, I'll surely go to Heaven after I die." Let me tell you that you won't be able to go there. Why not? Think about it. What is true faith? You're merely saying with your mouth that you have faith, but in your heart you don't actually have faith. Why do I say that? Because when you truly have faith your actions match your words. When you confess in church, saying, "My Lord, please help me. I did something wrong today. Please forgive my sin," your conduct is rather good at that moment. But once you step out the church door, you do as you please and behave just like an everyday person. You make the same mistakes, and among everyday people you even behave worse than them. Then think for a moment: What was the use of your confession? Many people don't go deeper than what's on the surface. You shouldn't repeat your mistakes after confessing. When you later find that you still make other mistakes, you go again to confess and stop repeating the same mistakes. As you do this and continually break from the mistakes committed among everyday people, aren't you a better person? Haven't you reached a higher realm and become a good person who far exceeds everyday people?! Aren't you cultivating and elevating your xinxing? Since Western culture doesn't have the concept of self-cultivation, neither Jesus nor Yahweh mentioned the term "cultivation" back when they taught their Fa; there wasn't even that concept. The reason is, they knew "Cultivation depends on oneself, while gong depends on one's master." As long as one practiced cultivation, everything would be helped by the god of the heavenly paradise one would go to. He would effect changes in your body, transform your body, and enable your body to attain the great power that a god has. All of that would be done by gods.

Some people say that they'll go to Heaven after they die. People who haven't practiced cultivation are, in the words of your Western culture, called "spirits" after they die. These spirits are actually ghosts. Think about it, everyone: What are they? People in the East have a clear and definite concept of human body cultivation: One develops a Buddha-body through cultivation, that is, a divine body. If a person doesn't have this Buddha-body—in other words, if a person doesn't have the body of a heavenly paradise's god—his spirit fears light after coming forth [from his body]. It disintegrates in sunlight, and can't even reincarnate. That's why people are covered with a cloth after they die—it prevents exposure to sunlight—and buried in the earth so that their spirits won't disintegrate. Buddhas and gods are infinitely brilliant, even brighter than sunlight. Wouldn't a person's spirit dissolve if he were to see a god? How could it go to Heaven? It thus needs to have the body of a god and, at the same time, change the compositional elements of its intelligent beings through cultivation to make them comprised of high-energy matter. Only with a Buddha-body or a god-body can it go. Can a spirit come to the human world? No, it can't. So before it can stay in this world, it needs to have a human body to reincarnate with and needs to be born through the bodies of parents. Some people mistakenly think that they'll go to Heaven if they just have faith. How could that be? It sounds ridiculous to those of us in cultivation circles. You have to truly cultivate before you can go to Heaven.

So in Western religions, in the process of having faith in God, a person's xinxing and his realm of mind continually elevate, and he becomes a better and better person. Then wouldn't you say he's cultivating? Form is unimportant in cultivation; there are all kinds of cultivation methods. It's a difference in cultivation form between Western and Eastern religions. Why don't they mention cultivation in the West, like they do in the East? Both are in fact cultivation. It's simply that Eastern culture is very specific with regard to this, while Western culture doesn't talk about these things. That's because when gods come down to save people and to impart their Fa, they need to adopt a cultural form that's intelligible to people. Only then can you understand—only [when they] use the human languages of the time can you understand. What I've said is also to tell you that practicing ordinary qigong can't enable people to reach a very high, profound realm. That's impossible. Only real cultivation can achieve that. Qigong is only something at the lowest level in cultivation, so it's limited to healing illness and maintaining health. But even when it comes to healing illness and maintaining health there are some requirements for people, because after all, [qigong] is a form of cultivation and is different from any kind of medical treatment that exists among everyday people. If you want to learn qigong to heal your illness, you have to meet the requirements of qigong's principles." (Teachings at the Conference in Europe, Li Hongzhi, May 30–31, 1998 ~ Frankfurt, Germany)

Student*: "Is Falun Gong consistent with Christianity in any way in its view of the development of humankind and the earth?"*

Master Li:

"Of course, I won't blame new students for not having studied the Fa in depth. What I'm teaching

are the principles of the cosmos. And this Dafa is the grandest, it penetrates the cosmos from top to bottom, and it can create living environments for all beings of the cosmos as well as for gods at different levels and humans. How could there be any comparison? This Dafa of the cosmos has created living environments and conditions at different levels for living beings of different levels. In other words, the living beings at a given level have to conform to the principles at that level and to the Fa of the cosmos at that level in order to exist there. To put it even more simply, the future environment in which you will live is determined by the height of the level you can cultivate to. The principles that Jesus stated in his time, however, were principles that taught the everyday people of society how to cultivate in order to go to his paradise. Where is his paradise? To be precise, it's within a small expanse of our Milky Way, in a celestial body corresponding to our Milky Way. Jesus is at the same level as the conventional Buddhas mentioned in the East, whereas the Lord referred to by Caucasians in the West far exceeds that level. But neither of them has gone beyond this small universe. Yet our Dafa has created boundless and countless universes like this one, more than all the world's grains of sand. So would you say they can be compared? The principles that Jesus spoke about are only the Fa at the level of Tathagata in a small universe within the boundless Great Fa. That's the relationship.

Jesus is a very fine divine being. It's not just me who says that: Gods in the heavens all regard him that way. In order to save people he was crucified by humans—have you ever thought about what that means? And yet Christianity takes the cross as the symbol of God, the sign of the heavenly kingdom, and the symbol of Christ. That's the biggest disrespect to the divine. That is man's own doing, definitely not what gods have told people to do. Because the cross is placed on graves, it truly represents death—not a divinity. All gods in the universe condemn humans for doing that. Furthermore, Christianity has used the image of Jesus suffering on the cross as its symbol. That is something gods cannot tolerate—they absolutely can't tolerate it! But people are still doing it. People have been doing disrespectful things like that to gods for hundreds of years.

Jesus has many magnificent images. He performed many miracles among ordinary humans. And yet instead of taking his glorious and magnificent images as symbols, you have taken the crucifix as the symbol. Are you respecting this divinity or hating him—do you want him to be crucified forever? People, you don't know what you humans are doing. One day I was watching the news on TV in the U.S. when I saw two glassblowers in Taiwan who were said to be a famous couple. They consider Buddha statues a form of artwork, and have made a lot of glass Buddha figurines. Taking Buddha to be artwork is in itself blaspheming Buddha. But what's more, the figurines of Buddha that they made were missing torsos, with only a head and half a face. They made Buddha into Modernist and Impressionist stuff. But think about it: Those are magnificent and mighty gods. In the old days, when people mentioned God or a Tathagata Buddha, they felt a sense of reverence. Would they dare to treat them like that? You think that you're portraying Buddha, promoting Buddhism, and doing good things. But this is not just your average discourtesy: It is the greatest disrespect. You are actually cursing and slandering Buddha—distorting his mighty image. You're doing the worst thing. The other day I said that people's values have been inverted. Human beings just don't understand what I mean. They can no longer distinguish good from evil, righteous from wicked, or good from bad. All values have been turned upside down." (Teachings at the Conference in Europe, Li Hongzhi, May 30–31, 1998 ~ Frankfurt, Germany)

V. Jesus Christ (By Li Hongzhi)

"The human world is extremely complex. Just looking at a person here on Earth, there is no telling which dimension he is from. The divinity in that dimension still wants to have charge over him. What matters is whether this individual can discern; if he can, then he will be able to return [to where he came from]. But [that divinity] may know that this human being is no longer up to the task, so he can do no more for him. He doesn't want to let him go, however. Back when Jesus came forth to save people, [these divinities] thought that he was meddling in their affairs. "This person is mine, and it should be to my place that he is saved to. Why have you encroached upon our domain?" They wouldn't stand for it. It was because of such a mindset [that they acted as they did]. If you think about it, it's not right. Jesus didn't care who belonged to whom. Saving people was the goal of his coming. He saw that all beings were suffering and wished to give them salvation, to allow them to ascend. So as it was, Jesus affected the many divinities of a variety of dimensions. Ultimately their disagreements intensified to such a tremendous extent that they manifested in the human world, just as human conflicts might, and the entire brunt of it landed upon Jesus. Jesus could not free himself of it, with the only option being death—he was thus crucified, thereby dissolving the enmity that existed between them. With Jesus' ordinary body of flesh gone, there was

no Jesus there to make enemies with, and that put an end to his countless troubles. That is why it's said Jesus gave his life for all of humanity. That is the idea.

You know that when Jesus arrived among ordinary men to save people, he had to suffer for people, and he was eventually crucified. You don't know what this means. You only know that Judaism did that terrible thing. Actually, it's not that simple. There were other factors involved. If we're to follow the logic that you just expressed, think about it: Since Jesus had such great abilities and his father was Lord of the Caucasians, couldn't he do whatever he wanted? Why did he spend so much effort on Earth persuading people to do good, telling people principles of the truth, and suffering so much for human beings? And at that time he also had to endure suppression and sabotage from other religions. Why then bother to do things the way he did? It means that even though he had such great abilities, he still had to rectify the human heart in order to save a person. If any of you sitting here can't cultivate your xinxing to the point of Consummation, then I can't do anything about it either. I can teach you the principles of the Fa, I can help you evolve gong, I can eliminate a portion of your karma for you, I can protect you, and I can do a huge range of things for you that you don't know about but that are necessary when you Consummate and afterwards. But if your heart doesn't change and your mind doesn't elevate, then all of that is in vain. That's what I mean when I say that if a person's heart doesn't change, even a Buddha can't help him."

"Not being able to make a person improve doesn't mean nothing can be done when he's become bad. How could that be the case? He can be cast down to lower levels or exterminated at will. But since divine beings are compassionate... The "loving heart" that Jesus spoke of was in no way the love among humans. Taking human love for the "love" or "loving heart" spoken of by Jesus is equivalent to abusing Jesus' words, blaspheming Jesus, and blaspheming gods. The loving heart that Jesus spoke of is "compassion," only he used the language of Westerners to express it. At that time Western culture didn't have the concept of compassion, nor did it have a term for compassion. This is a disparity caused by the differences in time periods and the cultures of different regions. I think I've discussed all facets of this matter; that is, how everyday people handle these things, how to handle them during cultivation, and how to handle them after you eventually succeed in cultivation." (Zhuan Falun Volume II, Li Hongzhi, 1995)

Students: "*I am a Western practitioner, and I feel that in Zhuan Falun, "love" seems to have lost its status.*"Master Li:

> *"Let me tell you, the love that Jesus, Saint Mary, and Yahweh talked about is not the love that humans understand, but compassion. It's just that the human race then didn't have the word "compassion" in Western languages, so later people used the word "love" to replace the words spoken by God. This love is filthy to God, and compassion is sacred. Because of love, people indulge in sexual desire; because of love, people indulge in sinful promiscuity; because of love, people will do many bad things, things that God cannot tolerate. Love has a side that humans think is beautiful, which is to help each other, but it also has a negative side, while compassion is completely good. So the love that Western gods talked about in the past, I tell you clearly again, is called "compassion", not the love that people in the world talk about. It's because there was no concept*

of compassion in your culture. Your ancient people just recorded the words spoken by God using the term "love"." (Lecture at the Eastern Fa Conference in the United States, Li Hongzhi, March 27-28, 1999, in New York)

Students : *"For Westerners, "entity" is very strange and difficult to understand. How can we explain "entity" to Westerners?"*

Master Li:

"The six realms of reincarnation also exist in Western society, and for people, this is something they cannot escape. It's just that Jesus only talked about hell and didn't talk about reincarnation. Jesus said that people are sinful, but he didn't say that sin is because people have karma. This is because people's culture is different. It's not that there is no entity in Western culture. Do you remember that in the New Testament, there is an incident where Jesus exorcised a demon that had possessed a man? It is recorded in the Bible. And that demon said: "This is none of your business, Jesus. It's none of your business!" But Jesus insisted on casting it out. From this point of view, you don't fully understand your own culture." (Lecture at the Eastern U.S. Fa Conference, Li Hongzhi March 27-28, 1999, in New York)

VI. Extraterrestrial Life (By Li Hongzhi)

Student: *"Is there any way to communicate with intelligent beings beyond Earth?"*

Master Li:

"It looks like your attachment is none too small. Many of the questions raised today are by those

who have come for the first time. Having seen many movies, having read many imaginative stories, and having been influenced by a lot of things in society, people have developed a warped mindset that indulges in fantasy. You asked how to communicate with intelligent beings. Let me tell you all that an ordinary person can't do that sort of thing—he lacks the ability. If you could indeed communicate with intelligent beings but weren't a cultivator yourself and were without energy, you would be in danger. You might lose your life any moment. Besides, you would see gruesome spirits but no high-level beings, for your level would be too low and limited to only that of humans. So it would be a terrifying situation. It's even more terrifying if you want to communicate with living beings beyond Earth. Lives beyond Earth are just lives on other planets within this physical dimension that our human eyes can perceive. Yet those lives don't regard people as human beings. They think people are beasts, a kind of animal. They slaughter people at will and experiment on you at will. They kidnap people to their planets, lock them in cages, and put them on display as animals. Many of Earth's people who have gone missing were taken by them.

I can also tell you that today's human science was created by them. Why have those aliens created for humans something like the current science? It's because what they have on their planet is akin to it. They want to bring it to this planet and eventually replace humans. They found that the human body is really perfect, flawless. So they covet the human body and want to eventually replace humans. They have added their things into one layer of human beings' molecular cells. That's already been done now and it has been formed on a large scale. So for me to save you today, I not only have to remove all the attachments and karma that you need to eliminate, but I also have to clean up this stuff for you. The road humankind is traveling down is dreadful. Computers were [meant] to one day command people—that's the direction as development continues. Humans were to become the slaves of computers and machinery, and eventually to be replaced by aliens. Why have computers developed so quickly, and why has the human brain suddenly become so "sharp"? All of this has been carried out by aliens controlling the human mind. Aliens have registered everyone who knows how to operate a computer. For real. As to our students, I have cleaned all of that up for them so that they won't be interfered with by aliens when they use computers.

Why are there aliens? Some aliens originate from their planets. Why do some aliens always come to Earth? It's because they were once humans on Earth. But it wasn't the Earth of our time. It was the previous Earth that existed at the same location as this Earth. Earth has been replaced many times. Each time it was replaced, there was a portion of lives—living beings created by gods with different appearances, some of whom were similar to today's humans, some of whom had different forms—who were relatively good at the time and were thus preserved. They weren't destroyed, yet they weren't allowed to enter the period of human development on the next Earth, either. So they were sent to other planets. That's how it was done every time. Also, [this process] has left a record of this universe's past, for the aliens were humans on Earth in different time periods. [This process] has served that purpose. But due to a slackening in overseeing them, they have developed, and they have created all kinds of scientific methods according to their schemes; some are like the kinds on our earth, some assume other forms, and there are a variety of modes of development. Of course, they don't use the term "science." The names they use belong to their languages and concepts. These things have deteriorated further and further. And with the deviation of the cosmos's Great Fa, the aliens have become worse and worse too. That's why all aliens are in the process of being completely eliminated. They are being eliminated throughout the entire universe. At present, those up there have been eliminated. What's left are only those who have escaped to Earth, who are mingling among people or who have attached to human bodies. And there are other types too. I'm watching all of them. Let them do their outrageous things for the time being. It won't be long before they're eradicated. (Enthusiastic applause) The corruption of humankind is directly related to them, as well. The things they have done are sins, so they have to be eradicated." (Teachings at the Conference in Europe, Li Hongzhi, May 30–31, 1998 ~ Frankfurt, Germany)

VII. Artificial Intelligence (By Li Hongzhi)

Student*: "Is it true that human history is just the 7 billion lives reincarnating among themselves?"*

Master Li:

"Maybe it is. You're all interested in this topic... The fact is, in history there has never been a society like the one today. You may have read in the newspaper someone's article about my addressing the issue of aliens, and he found it to be laughable. Think about it, everyone: There has never been a human race like this in history, yet there have been periods of time in history when the technology of the human race was more advanced—far, far, more advanced than it is today. Human beings

nowadays can't build a moon and place it in the sky, yet in the history of the human race, human beings were able to do so. However, the science developed by the human race of today and the current circumstances of today are not the only path in the entire process of the human race's development. There are various paths of development. The science of today's human race was, in fact, brought over by extraterrestrials. Some people are studying the extraterrestrials and collecting their so-called signals. As a matter of fact, they are precisely right around you, only they don't communicate with you. Also, they are doing what they intend to do in a very systematic way.

Think about it, everyone: The other day, you all understood correctly when I mentioned that science is a religion. In religions, they have their founders and priests, while in the sciences there are also various titles: university president, doctor, master, bachelor, professor, assistant professor, and so on. And moreover, it's a perfect form of religion, one that exists everywhere and that is highly systematic. Human beings believe in it more than they believe in any religion, and this belief has come about imperceptibly. If you don't study it well, you will be cast out by this society: You won't find a good job or have a good future. Everyone knows that a regular religion has you believe in it mentally, and then it has you see, hear, and feel the actual, real existence of gods. In contrast, the religion of science has you view things and move forward on a material basis so that you will then be mentally dependent on it. It takes the opposite approach.

But I'm not opposed to science, for it too is a product of the universe. I'm only telling you what science is. Science is not scientific, and it has brought many a disaster to the human race that will never be corrected. You all know about air pollution. The air that industries have polluted can never be restored to its purest state. Industries have also polluted water, and no matter how good your methods are, you can never restore it to its cleanest state. Human beings breathe this air and drink this water. If it goes on like this, the human race (not including Dafa disciples) will become deformed. The structure of their limbs will be deformed and it will get worse and worse. It's a disaster brought about in the material realm. In the spiritual realm, before it progresses to that point, science will claim that you are superstitious and unscientific, and strike you with the club of science if you talk about anything that science has not yet recognized, or if you talk about the existence of gods or the fact that good is rewarded and evil punished. What does that kind of blow result in? It has destroyed man's belief in the divine and the concepts that sustain the moral values of human beings. And when these are destroyed, the moral standards that sustain the human race are utterly destroyed. So nowadays there is no bad thing people fear doing. Wasn't all this caused by a science that actually isn't that advanced?

I'm only telling you the truth. I'm not saying I am against science. It's only owing to human beings' having excessive blind faith in it that it has imperiled the human race. And what's more, this science was brought here by extraterrestrials. What is their goal, then? They systematically brought to the human race subjects such as mathematics, physics, and chemistry. Then in your body and mind, the frame of thinking that they wanted has formed. Nowadays, production and tools of labor, forms of work, daily necessities, as well as ways of life brought about by science have shaped man's ways of thinking and all of his behavior. This has fully formed the mindset that makes up warped modern man. Nowadays, almost everyone is like that. They all have thoughts shaped by extraterrestrials, and no one is able to get away from it since everything you encounter was brought about by modern science. It seizes every opportunity, so every part of your life is inseparable from science.

Consequently, your mind has formed a way of thinking and mentality based on an understanding of the material world—an outward appearance that modern science has created. It's also a layer of particles. Your behavior is dictated by your mind. Then, the things you operate, the vehicles you drive—all the things you do—were all brought about by science. Thus, a layer of particles was formed in both the human body and mind. The science that was brought to human beings by extraterrestrials has formed a layer of particles belonging to their science. This is exactly the systematic arrangement made by the extraterrestrials in order to acquire the bodies of human beings.

Thus, as you might know, science takes advantage of the desires people have, and human beings' desires drive them to progress as they do. Whenever someone raises doubts about today's science, it often brings opposition from scientists. Human beings don't realize that this kind of intolerance towards questioning is itself the result of science. It should be clear that the intolerance of these "scientists" arises out of emotion on their part, not reason. I can say that after having reached this point, the things invented and created by scientists are in fact not the creations of human beings. Whose creation are they, then? They were created by the structure that is controlled by extraterrestrials and that has formed in man's brain. And that structure is precisely something tightly controlled by the extraterrestrials. They will give you an inspiration and have you invent something. They have begun to replace the human race in a systematic manner.

Once that has happened, you know, people's desires then drive human development. At first people wanted to manufacture robots, but now it's not just robots. They want to make human beings—to clone human beings. If human beings are indeed created through cloning, think about it... Human beings are managed by gods. When a person is born, if he or she has no soul (yuanshen), then it is only a corpse. Why can a human being die when he has no problems with his physical body? How can he die suddenly? It's that his soul has left. Without that soul, it is only a hunk of flesh. With his soul, he is alive. A human body is like a piece of clothing: When it is worn by a human being, it is alive; if it's not worn, it is dead. Think about it, then: What will happen if gods don't place a soul inside a cloned being on account of its having been created by human beings and gods surely not acknowledging it? An extraterrestrial will fill the vacuum by placing itself as the soul of that human. So then it will have a human body and will occupy a human being. There will be more and more of these types of beings. And owing to their desires, human beings will be ordered to continuously make them. More and more of them will be made. These beings will become the majority, and they will become members of the human race, and they will be smarter than man. The minds of human beings already have a layer of their particles in them and are being controlled by them. So they will formulate laws that won't allow human beings to give birth from a certain point on. All [lives] will have to be cloned. And that is when they will invade the earth on a large scale. On the surface they would look like human beings, but would not be; they would be alien lives. Think about it, then: In no way am I spinning tall tales for people! (Applause)

Of course, human beings are completely helpless in the face of current circumstances. They can't break away from all that has been given to human beings by science, for that science is controlled by the extraterrestrials, and you can't even locate them. Human beings want to contact the extraterrestrials, but almost everything of human beings today was invented by them. Human beings think that they won't contact you, but in fact you are connected to the extraterrestrial lives all the time. Is there really a need for you to look for them? So human beings are helpless in the face of all this. They can't solve it. This is what I see as one of the biggest problems for the current human race. I have spoken out and told people now.

I said that I don't want to oppose so-called science, yet I have seen the true predicament of the human race. Since I am offering salvation to you in this setting, I have to live in this place, riding in cars and flying in airplanes. Science and the extraterrestrials are also products of the universe. However, the development of the human race will always have its beginning and will develop to its end. This is a law. All I have done is to tell you one of today's phenomena in order for my students to understand the universe as they cultivate. If I go into the details many ordinary people won't believe them, so I won't talk about them. The development of the computer is in fact a terrifying thing for man. You might destroy yourselves without even knowing how you were destroyed. These days people store data and then analyze the results with computers. With further development, people might use computers to make decisions, and that would signal the beginning of human beings coming to obey computers. Computers would then begin to totally command people. Computers would become more and more intelligent, and human beings would rely on computers ever more, until ultimately they were totally controlled by computers. This is how man's desires would be used to destroy the human race." (Teachings at the Conference in Canada, Li Hongzhi, May 23, 1999 ~ Toronto, Canada)

"The road humankind is traveling down is dreadful. Computers were [meant] to one day command people—that's the direction as development continues. Humans were to become the slaves of computers and machinery, and eventually to be replaced by aliens. Why have computers developed so quickly, and why has the human brain suddenly become so "sharp"? All of this has been carried out by aliens controlling the human mind. Aliens have registered everyone who knows how to operate a computer. For real. As to our students, I have cleaned all of that up for them so that they won't be interfered with by aliens when they use computers. "(Teachings at the Conference in Europe, Li Hongzhi, May 30–31, 1998 ~ Frankfurt, Germany)

VIII. Author's Words to Christians

I have never liked others telling me what to do in important matters that affect my life. I rarely believe in the definitions of something important to me given by experts and authorities. Because many of these so-called experts and authorities are just politicians wearing the mask of experts.

As an immigrant from communist China, I learned two most important life skills while studying in the United States: writing and critical thinking.

After a year of intensive study of Master Li Hongzhi's teachings and writings, I would like to talk about my personal understanding and conclusion of *Zhuan Falun* and other articles by Master Li Hongzhi from my own research, investigation, and analysis.

1. My conclusion is whatever the *Bible* told us was pretty much true but there's a little difference caused by the human factors involved.

The differences between the biblical stories and historical reality caused by human factors are intentional by the divine world. For instance, Master Li Hongzhi explained because the divine world did not allow Jesus to write the Bible, it had to be written by humans because humans make mistakes and cause the biblical verses to appear confusing and contradictory. In this way, God can see who still firmly believes when people are confused and contradictory. This can be confirmed from the story of Job in the Old Testament. Job was a rich man with a happy family, and he was very devout to God. One day Satan was trying to sow discord between Job and God. Satan told God that Job was devout to God because God gave Job everything, so God began to test Job's faith in Him. Job lost his wealth, his family, and even his personal health. Job was confused, he didn't understand why God treated him like this when he's so faithful to God. But Job did not complain to God, instead his faith in God was as solid as a rock. In the end, Job passed the test, and God gave everything back.

The Buddhist scriptures were also written by humans. Why didn't the Buddha write the Buddhist scriptures?

The Buddhist scriptures clearly record the Buddha once told his disciples that the Buddha's Dharma can only save people for 500 years because nothing lasts forever.

500 years after Buddha, Jesus came to the world. Until now, Jesus' teaching has saved mankind for more than 2000 years, and again nothing lasts forever.

I used to think Jesus would come back in His flesh again to save the world. But this Eastern saint, Li Hongzhi, made me understand it will be unfair for Jesus if He comes back in his flesh again.

First, in the eyes of God, the human world is just a cesspool, very filthy and wicked. No divine beings would enjoy swimming in a cesspool. He made me understand no matter what I do or say, I should always put myself into other people's shoes, even if the other party is the Lord. Being in the human flesh is not the natural state of being for Jesus. In fact, our physical bodies are not our natural state of being either.

Second, Jesus is always with us, but we cannot see Him with our eyes. The bodies of all higher life forms are composed of very small particles and atoms that are smaller than quarks and neutrinos. Then we should ask why prophet Brandon Biggs can see Jesus but most of us can't. Shouldn't we look for the problem within ourselves? Just like Master Li Hongzhi said, "If you always point your finger outward, it means you are looking for answers outside, that is the way of the devil. If you look inward, all the answers will be found within yourself, that is the way of God." Is our problem that we do not follow what Jesus said or that we simply do not understand the meaning of what Jesus said? So, I suggest if you feel confused when reading the *Bible*, you should read *Zhuan Falun* as a reference book, and you will be amazed to find all your questions about the *Bible* will be answered.

In short, one of Master Li Hongzhi's priorities is to reinforce Christianity by clearing up all the confusions and the contradictions in the *Bible*. He knows America is a country founded on Christianity. He knows by saving America, the world can be saved. He knows Jesus better than any human.

2. My personal understanding of Jesus' teaching.

When Jesus talked about big love, he was actually teaching us Truthfulness, Compassion, and Forbearance. Why?

First Jesus didn't teach us to lie, so that's truthfulness.

Second, when Jesus talked about love, He meant compassion. Because the language at the time did not have the term "compassion", He had to use love instead. However, Master Li Hongzhi pointed out love is selfish, but compassion is not.

Last, when Jesus said, "If someone slaps one side of your face, then give the other side of your face for him to slap" he was telling people to be forbearing. Forbearance can eliminate your sin and gain virtues. Virtues can be transformed into your superpowers. These superpowers are innate but are covered by our material desires. When you have enough virtues, your superpowers will be transformed and manifested, just like what Sadhguru, Sun Chulin and Brandon Biggs are capable of doing, including the superpower of freeing your soul out of your physical body, allowing you to communicate with higher life forms like Jesus and be on your way going back to your Celestial Home with Jesus.

CHAPTER 6:

The CCP Is the Bane of the World

I. The New York Times Is in Cahoots with the Nazis, Stalin, and the CCP

The Falun Gong group has been peacefully revealing the evil of the CCP to the world. Shen Yun is an art group founded by Falun Gong practitioners. Shen Yun uses exquisite traditional Chinese dance to show the charm and kindness of Chinese culture before communism.

Since August 2024, *The New York Times* has published articles against Shen Yun Art Troupe that align with the CCP's agenda. The newspaper has published at least nine similar offensive articles in under five months. This is not surprising; from the Nazis to Stalin to the CCP regime, *The New York Times* has a history of ignoring human rights and supporting totalitarian governments.

The New York Times' disregard for facts and blatantly selective smearing clearly deviates from the basic ethical standards that the media should have. Looking back at The New York Times' reporting history, this is an unsurprising practice.

During the Nazi Germany period, *The New York Times* downplayed the Holocaust, rarely published reports on the Holocaust on the front page, and related content was usually placed in an inconspicuous place in the newspaper.

Historians and commentators believe the newspaper failed to fully inform the public of the scale and urgency of the Holocaust and failed to use its influence to pressure the U.S. government or allies to take stronger actions to prevent genocide or prioritize rescuing victims, which is considered a media failure.

During the Soviet famine, Walter Duranty, a *New York Times* reporter who covered the topic, said the Soviet famine was "basically nonsense" and reports of millions of deaths were exaggerated. His reports often reflected the official statements of the Soviet government and praised Stalin's policies.

Commentators believe Duranty acted as an overseas propagandist for the Soviet regime, covering up its atrocities and shielding it from global supervision.

Now, *The New York Times* is distorting Falun Gong and Shen Yun in almost the same way, supporting the Chinese Communist regime's agenda.

Over the past 25 years, tens of thousands of Falun Gong practitioners have been persecuted in China. The CCP's persecution of Falun Gong includes forced labor, torture, and even live organ harvesting. These heinous crimes demand global attention, but *The New York Times* has little interest in investigating them.

In fact, as early as 2017, when *The Epoch Times* began serializing exposing the bloody history of Communism, exposing how Communism committed its atrocities, and what it still intends to do today, *The New York Times* went the other way. Instead of exposing the well-known genocide, red terror, and the hundreds of millions of abnormal deaths caused by this ideology under communist rule, the newspaper published a series of articles defending the communist "red century" and published articles with titles such as "*How Mao Molded Communism to Create a New China,*" which is directly derived from the CCP's propaganda slogan.

In 2018, *The New York Times* published a column titled "*Happy Birthday, Karl Marx. You Were Right!*" which stated, "We are destined to continue to quote his ideas and test his thoughts until the kind of society he worked to build and for which more and more of us now desire finally comes to pass."

What kind of newspaper would publish this? Marx's theories inspired the establishment of the most brutal regime in the world, and *The New York Times* praised the man.

Mark Levin, host of the Fox News program "Life, Liberty and Levin", explained the reason last November.

> *This is not new for The New York Times. But I would say that this is more than just some superficial agreement or some conventional political debate. In fact, I would say that The New York Times supports the Nazis, supports Stalin, and supports the Chinese Communist Party today–including the human rights violations committed by these regimes, because The New York Times seems to support these regimes at a fundamental ideological level. Ultimately, The New York Times openly supports communism, the foundation of all these totalitarian regimes.*

The *New York Times* Downplays the CCP's Human Rights Violations.

In this series of articles, the author deliberately downplayed the severe persecution of Falun Gong practitioners in communist China since 1999, describing the Chinese Communist Party's prison system's forced organ harvesting from living Falun Gong detainees as just a "narrative."

On the afternoon of June 25, 2024, the U.S. House of Representatives unanimously passed the Falun Gong Protection Act (H.R. 4132) by voice vote. The bill requires the United States to stop the CCP's state-backed campaign of live organ harvesting from Falun Gong practitioners and other prisoners of conscience and requires the United States to impose sanctions on those who participate in and assist in live organ harvesting in China. Sanctions include freezing property in the United States, prohibiting entry into the United States, and a series of economic and prison penalties. I would like to ask *The New York Times*: are the investigators who have objective evidence, the victims and their families, experts, and all members of Congress who passed this bill lying?

Now the world has recognized the true face of the CCP, there are still some elites in the business community and even members of the royal family who use the resources and influence they control to inexplicably speak for the CCP and support the CCP.

Sometimes I wonder why these people, who have plenty of money, are still speaking for the CCP. What did the CCP give them to make them so loyal that they still speak for the CCP? Is it possible that the CCP bribed these influential people in political and business circles with organ transplants?

An expert once told me some donated organs can only survive for three to five years in the recipient, so every three to five years, the recipient must receive a new organ. Will the CCP use organs to threaten these world elites to speak for them? This is terrifying when you think about it.

In 2016, when one of *The New York Times's* own reporters, Didi Kirsten Tatlow, attempted to investigate and report on the CCP's killing of Falun Gong practitioners for their organs, the newspaper immediately banned her.

"My impression is that the *New York Times*–my employer at the time–was not happy that I was doing these reports (about organ transplant abuse) and initially tolerated them," the journalist told an independent tribunal in China in 2019. "After my journalistic investigation was blocked, I was unable to continue." In 2019, China's Independent Tribunal was a panel of independent experts charged with reviewing evidence of forced organ harvesting.

The New York Times Attacks the Anti-CCP Art Group

The *New York Times*' slander of Shen Yun directly shows people who understand the CCP the CCP's common practice of distorting facts.

Over the years, the CCP has continuously threatened the Shen Yun Performing Arts with shootings, bombs, and even murder.

In 2023, the tires of two Shen Yun tour buses were damaged in Costa Mesa, California, which could have caused a serious accident.

According to pictures provided by Shen Yun and a Costa Mesa police report obtained by the *Epoch Times*, the tires of the two buses had their rubber cut in half. This damage would not cause the tires to deflate but would burst at high speeds.

Because similar vandalism incidents have occurred in the United States and Canada before, Shen Yun Performing Arts had to hire security personnel to guard the buses day and night. However, in Costa Mesa, the Shen Yun buses were sent to a local repair shop for repairs, and it was only after the buses were picked up that the tires were discovered to be damaged.

The training base for Shen Yun performers and the boarding school for students is located in Dragon Springs, New York, and is guarded by security. The *New York Times* used this as an excuse to attack Shen Yun for restricting the personal freedom of performers and students, which prevented students from going out at will. However, the *New York Times* deliberately ignored in all boarding schools in the United States, underage students are not allowed to leave the campus at will for security reasons. Moreover, the CCP's spies have been seen outside Dragon Springs, ready to harass, attack, or even kidnap the faculty and students in Dragon Springs.

On November 17, 2022, former FBI Director Christopher Wray told members of Congress he was deeply concerned about the Chinese government's establishment of unauthorized "police stations" in American cities to conduct potential infiltration operations, which violated U.S. sovereignty. In mid-September 2022, the non-governmental human rights organization Safeguard Defenders released a report,

"Overseas 110–China's Transnational Policing Out of Control," revealing the CCP has set up overseas police service stations in many countries around the world, one of which is located in the pro-communist overseas Chinese group "Changle Association" in Chinatown, Manhattan, New York City. Compared with the concerns of the former FBI director, Shen Yun's faculty and students are in extreme fear every day. The school can only let students travel in groups to prevent them from encountering misfortune.

In sports and art performances, many athletes and artists often participate in competitions and performances with minor injuries due to their professional ethics. Which NBA player has not played with a minor injury? This kind of adherence to professional ethics was interpreted by the *New York Times* as Shen Yun abusing actors and allowing injured actors to participate in performances. According to Professor Zhang Tianliang of Feitian University, Shen Yun Performing Arts Company has never forced any actors or students with minor injuries to perform on stage. Whether it is an out-of-town performance or in school, there is a professional emergency medical team. If there is an incident like a sprained ankle, it will be sent to a local hospital for professional treatment and recovery.

The *New York Times* interviewed a small number of former Shen Yun performers. For such a large group, there are always dissatisfied people. There is no matter, person, or group in the world which everyone is satisfied with. Is there any group that is 100% satisfied? Even in such a large group as the Catholic Church and Christianity, there are still people who are dissatisfied and unhappy. Especially for those who left the group, they must find a reason for themselves. They will not say they are wrong, but instead the group is wrong and that is why they left.

However, this cannot be used as a reason for the *New York Times* to make up stories based on these interviews and deliberately attack Shen Yun and Falun Gong.

The New York Times did not condemn the CCP's persecution but instead attacked the victims of CCP's persecution-Falun Gong and Shen Yun, which have grown stronger amidst adversity and led people to return to "Truthfulness, Compassion, and Forbearance". *The New York Times* is on the side of the CCP.

II. The Culprit of All Man-made Global Catastrophes-CCP

The Book of Revelation 17:1-18 in the Bible mentions a great whore of Babylon riding a red dragon who committed adultery with the kings of the world and brought disaster to the world.

Doesn't this sound like the CCP is using economic means and female spies to seduce officials from various countries to make them corrupt and infiltrate government agencies in various countries, making the politicians speak for the CCP, and work for the CCP? And, in return, these corrupted politicians use unfair trade to continuously transfuse blood and prolong the life of the CCP dragon. This is how they realize their ambition to rule the world. This is why the CCP, the great prostitute mentioned in the Book of Revelation, has not perished yet and is constantly causing disasters in the world.

How many of the world disasters I have listed below do you think have nothing to do with the CCP?

- 2023 Hamas massacre in Israel (Middle East war)
- 2022 Russian invasion of Ukraine (European war)

- 2019 COVID-19 outbreak (biological warfare)
- North Korea and Iran's nuclear capabilities
- Trade wars against other countries (Deindustrialization of the United States and Europe)
- Violation of intellectual property rights of various countries
- National debt trap (One Belt One Road)
- Theft of commercial and military secrets of various countries
- Unilaterally tearing up international treaties (Hong Kong)
- Suppression of Hong Kong's anti-extradition movement
- Forced organ harvesting
- Genocide (Xinjiang concentration camps)
- Fentanyl smuggling (Unrestricted Warfare against the U.S.)
- Territorial disputes with the Philippines in the South China Sea
- Deliberate murder of Indian soldiers on the Sino-Indian border
- Incitement of nationalism leading to the killing of a 10-year-old Japanese boy
- Future invasion of Taiwan (World War III)

The crimes committed against all humanity by the CCP are too numerous to list.

You may think the CCP has nothing to do with you because the bad things the CCP has done have not affected your life. OK then. Let's talk about the CCP's persecution of you.

Illegal Immigrants Vs. the United States

Those illegal immigrants who come to the United States are victims of the CCP, because most of them come from dictatorial countries, such as Venezuela, and the CCP supports the dictators in their countries behind the scenes. These incompetent, selfish, and violent dictators have ruined the economy, causing people to live in poverty, social unrest, and even civil war. Do these illegal immigrants need to be supported by taxpayers? Afterward, most illegal immigrants will compete with you for job opportunities.

What's worse is there are many criminals and terrorists mixed in. A 12-year-old girl named Jocelyn Nungaray from Texas was raped and killed by two illegal immigrants.

Police said on June 20, 2024, two suspects were arrested, 21-year-old Johan Jose Martinez-Rangel and 26-year-old Franklin Pena. The two will face capital murder charges. The medical examiner has determined the victim was strangled. According to the Department of Homeland Security, the two suspects are illegal immigrants from Venezuela. They were caught at the border but later released by patrol officers. If you pay attention to the news, you may often hear news reports about illegal immigrants committing crimes in the United States.

I think the criminals who are among the illegal immigrants will not let you, your mother, wife, girlfriend, and daughter go just because of which party you voted for or did not vote for. Criminals have nothing to do with political correctness when choosing their victims.

I am not against immigration; I am an immigrant myself. And the United States is also a country of immigrants. I just want to know who these illegal immigrants are?

Dead Men's Bones in Your Mouth

Maybe you are a lucky white collar with a good income and the influx of illegal immigrants into this country doesn't affect you much, so okay, do you go to the dentist? Who doesn't need to have a filling, a denture, or an implant? So, what would you think if you knew that the allograft bone used in any of these dental treatments might come from the bones of the deceased in some Chinese crematoriums?

A scandal involving the theft and sale of corpses involving biomaterial companies, funeral homes, and hospitals in various places in August 2024 recently sparked discussion on the Chinese Intranet. The relevant Chinese authorities censored related news and discussions with unprecedented intensity. In addition to the usual rapid deletion of related news and closing of social media comment areas, this time even liking the news was rarely banned.

"A case that shocked me"

On August 7, Yi Shenghua, a partner at Beijing Yongzhe Law Firm, exposed a large-scale corpse theft case on his Weibo that he learned from an insider. He said the case made him "stunned."

> *A pharmaceutical company in Shanxi cooperated with funeral homes in many places to purchase corpses aged 20-60 who died of non-disease," he wrote in a post. "The purchased corpses will be roughly cut up, and some bones will be used to make biological tissue materials and sold to hospitals." According to the prosecution opinion that was later leaked online, the entities involved include Shanxi Aorui Biomaterials Company, Sichuan Hengpu Technology Co., Ltd., Shandong Qingdao University Affiliated Hospital, Guilin*

Medical College, at least three funeral homes at the city and county levels in Guilin, and four crematoriums in Yunnan, Chongqing, Guizhou, and Sichuan. The main company involved, Shanxi Aorui, has an official background. The company was founded 25 years ago by the China Radiation Protection Research Institute. The indictment states that between 2015 and 2023, Aorui illegally purchased at least thousands of corpses from Sichuan, Guangxi, Shandong, and other places, and carried out "rough dismemberment, removal of flesh, cleaning, irradiation, etc." to make bone implant materials and sell them to hospitals in various places. Aorui made a profit of 380 million yuan, or about 53,748,232 US dollars, by reselling corpses in these eight years.

Many companies like Shanxi Aorui have official backgrounds and engage in such dark transactions. This is just the tip of the iceberg. In the current environment where almost all products are made in China, I can only pray the tooth powder used to fill my teeth does not come from the bones of the deceased of a crematorium in China.

Mind-controlled Weapons Threaten US National Security

On February 17, 2025, a shocking transaction shocked the Ethereum community. A Chinese programmer named Hu Lezhi burned 603 $ETH Ether ($1.65 million) and donated 1,950 $ETH ($5.35 million) to the Ethereum Foundation and Wikipedia, which has attracted the world's attention. Hu also sent a message accusing Feng Xin and Xu Yuzhi, CEOs of Kuande Investment, of using brain-computer weapons to harm all employees, former employees, and even themselves as controlled persons.

Later, Hu Lezhi published a statement on brain-computer weapons and digital slaves. He warned, "*A new type of crime model has emerged, in which the desires and perceptions of the victims are gradually deprived. Eventually, they become complete slaves of digital machines. If I become a victim in the final stage, I will choose to end my life.*"

Hu also stated that brain-computer chips have been widely deployed in the military field and used base stations, radios and nano brain-machine chips to control all Chinese citizens.

In addition, CCTV 7, the CCP's official media "Liberation Army Daily" and Xinhuanet, as well as some academic papers, have repeatedly confirmed the existence of brain-controlled weapons, and the CCP has been studying them for decades.

On December 22, 2021, U.S. Secretary of Commerce Raimondo pointed out that scientific exploration of biotechnology and medical innovation can save lives. Unfortunately, China uses these technologies to control people and suppress ethnic minorities and religious groups. On December 16 of the same year, the U.S. Department of Commerce sanctioned 34 Chinese companies and institutions, restricting the export of U.S. technology and products to these entities. In explaining the reasons for the sanctions, the United States accused the CCP of using "brain control weapons" to infringe on human rights. In recent years, more and more Chinese people have disclosed in their real names that they have been "brain controlled" and suffered extreme physical and mental harm.

With the advancement of the CCP's brain control technology, it can now control almost all Chinese people.

Hu Lezhi is comparable to Dr. Li Wenliang, the whistleblower who exposed the CCP's Wuhan virus in 2019. This also reminds me of the Wuhan coronavirus that broke out in 2019 and killed millions of Americans. It was also made as a bioweapon in the CCP's Wuhan laboratory. Other countries were studying how to weaken the lethality of the virus, while the CCP's studied how to increase the transmissibility and lethality of the virus as a biological weapon.

Today, for the same brain-computer interface technology, Elon Musk's Neuralink implants wafer into the human brain to treat paralysis and Alzheimer's disease, help those who cannot live a normal life due to accidents and diseases and restore their autonomy and quality of life. Imagine a paralyzed person who can control a wheelchair or even a computer through a direct access to the wafer in the brain. This science fiction is becoming reality.

However, the CCP's development of brain-computer interface technology is set out for brain control weapons. This may be the beginning of another world-class disaster similar to the CCP Wuhan virus.

Hu Lezhi said that Kuande Investment uses this technology not to cure diseases and save lives like Neuralink, but to control people and turn them into digital slaves. IT WANTS EMPLOYEES TO BE LIKE ROBOTS, COMPLETELY OBEY THE BOSS'S ORDERS, AND MAKE PEOPLE COMPLETELY LOSE THEIR OWN THOUGHTS AND WILL.

According to incomplete public information, the CCP has been researching "brain control" technology for at least 60 years, with the military, scientific research departments and universities all involved. However, in recent years, more and more Chinese people have revealed their real names as victims of "brain control" experiments, being harassed 24 hours a day by "intracranial voice transmission" and other means, suffering great physical and mental harm. Yao Duojie, 53, is one of the digital slaves or digital "killers" that Hu Lezhi mentioned.

In his article *My Life No Longer Belongs to Me*, Yao Duojie reveals that how he was selected as a live test subject for secret brain-control weapons since 1976.

His article mentions that when his grandfather passed away in 1985, he was heartbroken, but his facial expression suddenly became uncontrollable, and he was laughing.

But what made him even more uneasy was that in 1995, when he and his brother went to Hainan for a trip, he had an evil impulse to pick up his brother and throw him into the sea while standing on the rear deck of the cruise ship. He was extremely frightened at the time, holding the steel handrail tightly with both hands. Fearing that he could not control himself, he hurriedly asked his brother to return to the cabin.

Since then, Yao Duojie has been collecting information about brain control research. He found the book called *The Secret of Bio-electromagnetic Waves–Discovery of Field Guidance* (生物电磁波揭秘——场导发现) written by Jiang Kanzheng (姜堪政)and Yuan Xinzhou(袁心洲). The book introduced that in 1957, biologist Jiang Kanzheng (姜堪政) proposed the basic view of field guidance; in 1963, he assembled a "bio-wave amplification device" with simple electronic components. The instrument amplified A's human bio-waves and transmitted them to B, so that B could know A's thoughts synchronously. Jiang Kanzheng (姜堪政) asked his classmates to cooperate in field guidance experiments at that time, and the success rate reached 90%; in 1967, the Chinese Academy of Sciences (中国科学院 or 中科院) conducted secret research and development of human brain field guidance on the theory of bio-field guidance.

Yao Duojie also read *Neuroengineering* (神经工程学), which introduced how to study mind-controlled weapons, including collecting bio-electric waves generated by humans under various stimulation conditions as a database for brain-controlled weapons and human brain development.

What Yao experienced was a brain control experiment of mind reading and behavior control. This technology modulates the control signal on microwaves to generate induced currents on the nerve fibers of the target's brain. The control information carried in the current can directly affect the target's thinking and behavior. There are many such examples, too many to count.

In recent years, Yao Duojie and other victims have been handing out flyers on the streets, trying to educate more people about the "brain control" technology and telling the victims' painful experiences. Yao realized that the most painful thing about being controlled by the brain control technology is the passive thinking of the brain.

"When you see a chicken, it will make you think about how to cook it after I kill it? Should I braise it or stew it? When you see an old woman or a beautiful girl, it will input obscene things into you."

With cross-border underground tunnels and the open border policy during Biden administration, would the CCP import military personnel who are adept at using mind control weapons into the United States, and then mind-control American civilians and manipulate them to do illegal things? If, as Yao said, "the most painful thing about being brain controlled is that the brain thinks passively. When you see a chicken, it will make you think about how to cook this chicken after I kill it? When you see an old woman or a beautiful girl, it will input obscene things into you." Then, if a person who is brain-controlled and has a gun sees a school, what will that armed person do?

School shootings are so common in this country. Could this be related to the CCP's brain control experts who have infiltrated and been lurking in the United States?

Will American society become more violent?

If the CCP's brain-control weapons had made Yao wanted to throw his brother into the sea when they were on the deck of the cruise ship, then if this brain-control weapon is used in the United States to attack American citizens, the consequences will be disastrous. Imagine, during festivals or other large-scale celebrations, there will always be a large crowd of people gathering. If a nearby driver is controlled by a brain-control weapon, will he drive a truck into the crowd?

Will important officials of the US government and military be beheaded?

Because TikTok, a CCP spyware, has collected a lot of information about Americans. These TikTok users whose personal information has been collected are more likely to become victims of the CCP's brain-control weapons and be controlled by them. If the family members, friends, neighbors or colleagues of these brain-controlled Americans happen to hold important positions in the government and the military. Then in extreme cases, the CCP may assassinate important officials through these brain-controlled American TikTokers!

The CCP could easily import brain-control weapons, biological and chemical weapons, and even miniaturized nuclear bombs into the United States from the tunnels under the southern border.

The CCP would definitely not miss the opportunity to deploy a large number of spies and military personnel to the United States under the open border policy during the Biden administration.

The CCP owns a large amount of land and real estate in the United States. Much of the land owned by the CCP are next to US military bases. I think the CCP may have bought a lot of real estate in communities near the homes of many US government officials. A radio base station for brain-control weapons can be accommodated in the garage of an ordinary family or middle-sized commercial vans.

I think the worst-case scenario is that before Xi Jinping invades Taiwan, the water source of the US military base will be poisoned, many chip factories in the United States will catch fire inexplicably, many terrorist attacks will occur simultaneously in the United States.

III. World War III Is Not a Threat, but a Warning.

U.S. intelligence agencies have always attached great importance to the personality analysis of target persons. The study of foreign leaders has a long history. The first foreigner they studied was Adolf Hitler, the head of Nazi Germany. The researchers successfully predicted that Hitler would choose suicide after suffering a certain degree of failure.

According to a U.S. State Department cable leaked by WikiLeaks, the United States conducted a personality assessment on Xi Jinping, who was still China's vice president, as early as 2009. There have been many different opinions in the international community about whether Beijing will launch a war against Taiwan. Some people think it will, while others think it will not. But the majority of people analyze Xi Jinping from a rational perspective of pros and cons. Few people judge Xi Jinping's behavior based on the personality traits of a dictator. Just like before the Ukrainian-Russian war, many people analyzed and judged from a rational perspective that Putin should not launch a war against Ukraine. However, the opposite happened. Putin still launched the war knowing the Americans had predicted it.

The Psychological Characteristics of a Dictator?

In 1939, Carl Jung met with Hitler and Mussolini in Berlin and observed them. In Jung's memory, Hitler never smiled and always looked unhappy. He seemed to be sulking. Jung judged Hitler was sex-indifferent and cruel and had no humanity. Hitler had only one goal in life, which was to establish the Third Reich. He wanted to establish a powerful German Empire to eliminate the anxiety in his heart and wash away the shame Germany had suffered in history. Hitler frightened Jung.

Jung's evaluation of Mussolini was completely different. He said Mussolini looked like an ordinary person, friendly and energetic. Later, history proved Hitler brought disaster to the world, while Mussolini contributed jokes to the world.

Psychology has now summarized six personality defects of dictators including Hitler, Saddam, Kim Jong-un, Putin, and Xi Jinping–sadistic tendencies, antisociality, paranoia, excessive love, schizophrenia, and split personality.

Some people may have such doubts, since these people have personality disorders and are mentally ill, why can they hold such great power? It is worth pointing out although these people have personality disorders, and mental illness, and are crazy or even perverted in some aspects, their IQs are not low. Some perverted mentally ill people have very high IQs, such as Charles Manson, the notorious serial killer in the United States. He was a serial killer, but at the same time, he had very strong leadership. He fooled a group of people into committing crimes with him and was also crazy about some women. The conclusion of psychiatrists is these people have strong leadership and the ability to deceive people, such as Stalin and Mao, and there are still many people worshipping them today.

Psychologists have concluded paranoia exaggerates the illusion and can cause schizophrenia to have violent tendencies. In fact, many people in our lives have these six personality defects which dictators have, but they have not become serial killers or dictators. Instead, they abuse cats at home and abuse their wives and children.

One thing which needs to be pointed out in the study of dictator personality is two parts of our brain are related to physiological satisfaction, one is the frontal lobe, and the other is the amygdala. When faced with moral decisions and impulse control, the frontal lobe activates the amygdala to regulate anger, fear, and desire. Both contribute to the brain's reward system. So, when these two parts of the brain are damaged or underdeveloped, it is the secret to mental illness. Things which satisfy a normal person, such as reading a good book, watching a good movie, or eating a good meal, can make a normal person satisfied. But these things cannot satisfy people with underdeveloped amygdala.

People need satisfaction, and for some people, alcoholism and drug addiction can satisfy them, while others can satisfy them by killing and abusing others, so they become serial killers. For dictators, only power can satisfy their insatiable desires, and this desire will gradually deteriorate. That is, as time goes by, they will become more and more vicious and cruel. Just like drug addicts, their addiction will become more and more serious.

Most people pursue power for enjoyment. After enjoying rights far beyond those of ordinary people, they are satisfied and start to become conservative. They shrink in the power struggle, but dictators are completely different. Since they are not easily satisfied by ordinary people's things, they must have a lasting motivation to pursue things to satisfy them. Look at Stalin and Mao Zedong. They also encountered many setbacks in their lives, but in the game of power struggle, they were more excited and crueler than their comrades. In the end, they won the battle to suppress opposition forces within the party and gained the highest power. Putin and Xi Jinping also defeated dissidents and gained the highest power because of these personality defects.

Dictators in the Eyes of Social Philosophy

Not only are psychologists studying dictators, but so are sociologists. Adorno, a famous social philosophy school of the Frankfurt School, analyzed the personality traits of dictators. He summarized the following seven traits of dictators:

First, there is a strong loyalty to certain habits, values, and ideals. Xi Jinping once told Biden that a dictatorship is better than a democracy. Therefore, Xi Jinping has always wanted China under his governance to replace the United States and re-establish his ideal international order.

Second, dictators have egocentrism centered on their own nation. This can be seen from Putin's frequent references to the power and brilliant culture of the Tsarist Russia in history, and his clear expression of his ideal of reviving the Tsarist Russia. Xi Jinping's most important political goal is the great rejuvenation of the Chinese nation. Due to the narrow perspective of the dictator, he will be hostile and even attack anyone who does not obey. The suppression of Hong Kong, Tibet, and Xinjiang, and the possible invasion of Taiwan in the future are the best examples.

Third, foreigners are seen as threats. The Communist Party's media uses the excuse that the United States and Western allies including Japan are suppressing China's national rejuvenation as an excuse to incite national hatred.

On June 10, 2024, a knife attack occoured in a park in Jilin Province, northeast China, and four American teachers from Cornell College in Iowa, USA, who worked in China, were stabbed.

On June 24, 2024, a Chinese man attacked a Japanese school bus with a knife, resulting in a pair of Japanese mother and son being injured and the death of a 54-year-old Chinese female school bus driver who bravely stepped forward to stop the crime.

On September 18, 2024, a 10-year-old Japanese boy was attacked and killed with a knife by a 44-year-old Chinese man outside the Japanese School in Shenzhen, China.

On October 1, 2024, a 23-year-old Chinese student stabbed three five-year-old boys on their way to a daycare center in Zurich, Switzerland. The three five-year-old boys were later confirmed to be out of danger.

On October 3, 2024, a 9-month-old Australian baby boy was splashed with hot coffee by a 33-year-old Chinese student, resulting in severe burns and seven operations. The incident shocked the whole of Australia.

In just four months, there were five incidents of Chinese citizens harming foreigners, which shows how much Xi Jinping and his supporters hate foreigners.

Fourth, he believes that he must be doing better than others. Xi Jinping believes only the CCP can maintain China's social stability and national strength. He also believes only the second generation of Communists like him are qualified to control the country their fathers fought for.

Fifth, aggressive leadership personality. When they are threatened or at a disadvantage, their attitude becomes terrible. Xi Jinping's purge of the Ministry of Foreign Affairs and the top brass of the CCP's military in 2024 can prove this.

Sixth, he has prejudice and stubborn ideas. During the outbreak of the CCP virus, Xi Jinping locked down such a large country for three years. The longer the lockdown, the worse the economy became, and the more he wanted to achieve dynamic zeroing. The more the United States regards China as its biggest enemy, the more Xi Jinping wants to vigorously develop its military. He wants to land on the moon and build a space station to attack the U.S. mainland

and its satellites. He even wants to change the rules of international law and international order, turning the South China Sea and the Taiwan Strait into China's inland seas.

Seventh, living in a personal world. Xi Jinping's predecessors only wanted to gain power and money for themselves and their families, but Xi Jinping is different. He is a nationalist with a strong belief in communism. He is determined to avenge the trauma and humiliation caused to China by the West for a century, and to make China an unchallengeable world superpower under his leadership. Does Xi Jinping's characteristics resemble Jung's evaluation of Hitler? This is why many people have nicknamed Xi Jinping "Xitler."

Xi Jinping is very old, and his time is running out, but his political legacy is very important to him, so at present, unifying Taiwan is the only thing that can temporarily satisfy this personality-defective dictator.

So, will Xi Jinping invade Taiwan?

My answer is that he will if he has the chance. Even if there is no chance, he will probably create one to invade Taiwan.

The CCP's massive dumping of goods around the world is to deindustrialize the United States, Europe, and other countries is the same method used by Nazi Germany before World War II. Hitler devalued the German mark and then sold cheap German industrial products to other industrial countries in Europe, causing many factories in other European countries to close down. After Hitler successfully deindustrialized other European powers, World War II was launched.

The industrial production capacity of various European powers, including Britain, could not keep up with the demand for war consumption due to the large number of factory closures, and they could not produce high-quality weapons to deal with the German army in time. This allowed the Nazis to quickly occupy a large area of Europe in a short period. After that, the Allies had to spend a long time, consume huge resources, and pay a heavy price in casualties to regain the land occupied by the German army.

Xi Jinping is now reducing product research and development costs by stealing intellectual property and lowering the wages of Chinese workers. Meanwhile, he has continued to devalue the RMB while selling many Chinese products to the United States and Europe. The political goal of deindustrializing the United States and Europe is on the way. What Xi Jinping is doing now is the same as what Hitler did before World War II. And yes, "Xitler" is preparing the war now.

CCP's Unrestricted Warfare against the United States

Unrestricted Warfare: (超限战 or 超限戰) is a book on military strategy written in 1999 by two colonels in the People's Liberation Army (PLA), Qiao Liang (乔良) and Wang Xiangsui (王湘穗).

The authors of this book try to combine the strategy of defeating the strong with the weak in Sun Tzu's *The Art of War* with the characteristics of modern warfare in the 21st century to defeat a technologically superior opponent (such as the United States) through a variety of means. Rather than focusing on direct military confrontation, this book instead examines a variety of other means such as psychological warfare, smuggling warfare, media warfare, drug warfare, network warfare, technological warfare, etc.

American generals are always full of confidence and assure Congress they will be able to defeat the CCP if it invades Taiwan. I am also deeply confident in our generals' expertise in defending our national interest and very grateful for their services. However, anyone who knows the CCP's war history knows none of the wars won by the CCP was in a gentlemanly way as we wished because the CCP's evil means are beyond your imagination.

IV. Ending the CCP in a Cost-effective Way

The main idea of *Unrestricted Warfare* is that today's wars are no longer mainly hot wars. With the emergence of the Internet, whether it is commercial theft against the enemy country, paralyzing the enemy's infrastructure, or destroying the enemy's economy through financial warfare, these are actually wars. In addition, there are drug wars, deindustrialization wars, and media propaganda wars against enemy countries. These wars are all carried out to destroy the enemy internally.

In fact, attacking the United States on American soil is a strategy formulated by Wang Huning for Xi Jinping. The Washington Post called Wang Huning Xi Jinping's chief Mastermind. The newspaper also called Wang Huning the most dangerous man in the world. In 1991, Wang Huning wrote a book called *America Against America.*

Wang Huning's philosophy is not only used in the United States, but also in South Korea and Taiwan. The CCP is now using South Korea to oppose South Korea and Taiwan to oppose Taiwan. At the end of 2024, the impeachment of South Korean President Yoon Seok-yeol by the South Korean Democratic Party controlled by the CCP has aroused widespread resentment among the South Korean people. The same is true in Taiwan. Now the Taiwanese people are launching a recall of legislators controlled by the CCP. So whether in the United States, South Korea or Taiwan, what Wang Huning is doing is to subvert these countries from the inside.

Facing this war that has already broken out on American soil, the United States needs a very pragmatic policy. It cannot just use defensive strategies, such as catching Chinese Communist spies. The counter-attack strategy must be offensive. Offense is the best defense.

"Invincibility lies in the defense, the possibility of victory in the attack"-Sun Tzu

Offense means to help the Chinese people fight for true freedom. Because when China becomes a free society, the threats that the United States and the world facing will no longer exist. This is a strategy of cutting off the source of the problem.

So, how can we help the Chinese people gain freedom? We all know that what the CCP fears most is its people. That is, we must support the Chinese people's armed revolution to overthrow the CCP. Because there are two kinds of terrorism in the world. One is called religious terrorism and the other is state terrorism. The CCP is the largest state terrorist group.

For terrorists, violence is the only language they understand!

Unrestricted warfare must be based on economic warfare and supplemented by truth warfare and proxy warfare.

1. <u>First, we must use economic warfare to destroy the CCP's finances</u>.

The first and most important factor is to look at the financial situation of a government. Even if a government does all kinds of bad things and is criticized by its people, it will not automatically collapse. A centralized government like the CCP will not be overthrown by being scolded or protesting. History has given the CCP many opportunities, but the CCP always chooses to suppress the people with violence. Whoever controls the army controls the situation, and good finance is essential to maintain an army or win a war. The victory of the American War of Independence was decisively affected by the loans and military aid from France and Spain.

Just like the American War of Independence, if the Chinese people want to gain freedom, they must overthrow the CCP through revolution. The state machinery used by the CCP to suppress the people includes the army, the Ministry of State Security, and the Ministry of Public Security. These require a large amount of financial expenditure from the CCP to maintain. Without financial support, these violent machines will stop working. This will bring opportunities for the Chinese revolution.

So how bad do the CCP's finances have to be to be considered conducive to a revolution? This requires observing whether the obedience of the military and police have been significantly reduced during the people's protests in the streets. The CCP authorities are having trouble commanding the soldiers sent to suppress the people. They are increasingly unable to control the situation through suppression and are increasingly relying on compromises to ease conflicts. In this case, the probability of a successful revolution is greatly increased.

2. Support the mature Chinese anti-communist revolutionary armed forces–V-Brigade

The Australian Broadcasting Corporation (ABC) Four Corners Show interviewed Eric Wang, a former double agent and one of the main founders of the anti-communist armed organization V-Brigade. According to ABC, from 2008 to 2023, he worked as an agent in the "Political Security Bureau" (abbreviated as "First Bureau") of the Ministry of Public Security of China and revealed how he was ordered to hunt down overseas dissidents, including pretending to be a real estate executive or even an anti-communist fighter to gain the trust of the hunted targets, and then luring them to certain countries for agents to kidnap them back to China.

The report said that the "First Bureau" operates around the world, mainly targeting people who criticize the CCP and Xi Jinping and conducting surveillance, kidnapping, and silencing. Eric described it as the "darkest department" of the CCP, which does whatever it wants against anti-communists.

Eric said he was attracted by the Western democratic system when he was in college in 2007 and joined the Chinese Social Democratic Party in the United States. Unexpectedly, he became a target of CCP surveillance. He was later detained by Chinese agents and forced to admit he attempted to overthrow the Chinese government, but was given a chance to "reform" by serving as an overseas secret policeman.

But Eric Wang is a national hero in the hearts of Chinese anti-communists who know about the V-Brigade. Eric Wang was one of the main founders of the anti-communist armed organization called V-Brigade. According to Eric's personal Chinese YouTube account v-brigade and his X account V_brigade_eric, when he was a former Chinese spy in Myanmar, he cleverly used the resources of the Chinese Political Security Bureau to establish the anti-communist armed V-Brigade and formed the Chinese Freedom Army with 11 other anti-communist armed forces on the Sino-Myanmar border. Australian intelligence agencies have obtained relevant information about the V-Brigade from Eric Wang and are currently investigating and confirming the existence of the V-Brigade.

The official spokesperson of the V-Brigade, Professor Yuan Hongbing 袁红冰), a famous legal scholar and editor-in-chief of the website "Fire of Liberty", has repeatedly given interviews to overseas Chinese media and detailed the two times the CCP launched an encirclement and suppression campaign against the V-Brigade in Myanmar, both of which were defeated by the brave fighting style of the V-Brigade soldiers. Detailed information about the V-Brigade can be found on the website of "Fire of Liberty" https://fireofliberty.info or its YouTube account 自由聖火Fire of Liberty.

The CCP has been using its propaganda tools in China and Western countries to cover up the existence of the V-Brigade. But after the CCP's second attempt to suppress the V-Brigade failed (starting on November 20, 2021, and ending on December 30, a period of nearly 40 days), it made three unusual moves:

First, all Chinese in Myanmar were called back.

Second, a 500-kilometer barbed wire fence was built along the China-Myanmar border.

Third, a minefield was laid along the 500-kilometer barbed wire fence built along the China-Myanmar border.

The CCP claimed it was to prevent telecom fraudsters from Myanmar from entering the country. However, it was interpreted by the outside world as preventing the Chinese Freedom Army led by the V-Brigade from infiltrating mainland China.

For more information, please contact the following:

V-Brigade leader Eric Wang's YouTube account: V-Brigade
Eric Wang's X account: V_brigade_eric
V-Brigade's X account: V_brigade
V-Brigade official spokesperson Yuan Hongbing（袁红冰）'s YouTube account: 自由聖火 Fire of Liberty
The "Fire of Liberty" website found at https://fireofliberty.info
China's Freedom Army's YouTube account: 中国自由军

Not just any revolutionary armed force can be competent to overthrow the CCP. Because the original intention of the revolution is to save people, not to kill people, to end the CCP, not for revenge killing. In the process of revolution, revolutionists must consider how to minimize the casualties of human life, including the lives of ordinary Chinese people, revolutionary members, and CCP officials and their families.

War is an art, not a science, and the unrestricted war against the CCP is no exception. It requires a revolutionary team that understands the CCP and is imaginative and creative to carry it out. So where can we find such a revolutionary team? The answer is simple. As long as we pray to God sincerely, the revolutionary team will definitely appear!

3. Support the spread of the truth in mainland China

An old Chinese saying goes,
He who wins the hearts of the people wins the world.

So, how to win the hearts of people like the Chinese, who are isolated by the Great Firewall and suffer from the enslavement and persecution of the CCP? Spread the truth about the CCP to the Chinese. The truth can induce indignation and injustice in the hearts of the victims. The indignation and injustice can be transformed into the courage to seek change, and the courage can be transformed into the action to seek change, that is, to resist or join the revolutionary army to overthrow the CCP.

As for telling the truth about the CCP to the Chinese people and even the whole world, Falun Gong has the most experience and has been doing it for the longest time.

Why has the CCP carried out an unprecedented and brutal persecution against Falun Gong for decades?

Why has the CCP never given up demonizing Falun Gong and its affiliated organizations?

Why is the CCP so determined to eliminate Falun Gong at any cost?

The CCP regime, which is maintained by lies and violence, almost perished because of Falun Dafa. This is the real history of China in the 1990s.

You can compare the CCP, who believes in falsehood, evil, and violence, to the bacteria and viruses on your hands, and Falun Dafa, who believes in truthfulness, compassion, and forbearance, is the hand sanitizer you use to kill these germs.

In the short seven years between 1992 and 1999, the social landscape of China and the overall moral level of the Chinese people were greatly changed by Falun Dafa.

The spread of Falun Dafa's principles of truthfulness, compassion, and forbearance rapidly improved Chinese people's moral standards, thus changing the social landscape of the communist state. People began to use the standard of truthfulness, compassion, and forbearance to measure their daily lives.

Under the influence of the concepts of Truthfulness, Compassion, and Forbearance, the quality of products and services provided by blue-collar workers not only met production requirements but also exceeded consumers' expectations. Officials at all levels of government no longer engaged in corruption and bribery, but instead truly served the people. Even schoolteachers began teaching traditional Chinese cultural virtues and abandoned the CCP's brainwashing communist ideology. This touched the CCP's cheese. We know the main reason why communism cannot be completely eradicated in the world is because it controls the military, police system, and most importantly the education system.

At that time, there were Falun Dafa practitioners who believed in Truthfulness, Compassion, and Forbearance in all walks of life, including the CCP's party, government, military, police system, and government agencies at all levels.

CCP's Politburo, the centralized power structure of the CCP, had seven standing committees which discussed and decided on national affairs. On July 20, 1999, Jiang Zemin began suppressing Falun Gong. However, the other six members of the Politburo Standing Committee did not agree. Jiang Zemin was the general secretary and the top leader and was extremely isolated in the suppression of Falun Gong. He did not expect the other six standing committee members to remain neutral to Falun Gong or stand on the side and speak for Falun Gong. Even the Premier of the State Council of the CCP, Zhu Rongji, told Jiang Zemin, "Let them practice," meaning Jiang should not interfere with Falun Gong's spread of the concepts of Truthfulness, Compassion, and Forbearance in China. Jiang Zemin had been driven crazy by his ever-growing jealousy. He immediately jumped up, pointed at Zhu Rongji's nose, and shouted, "Stupid! Stupid! Stupid! You're destroying the party and the country!" Zhu Rongji could not withstand Jiang's absolute power and gave up his support to Falun Gong.

Dafa is the natural enemy of the CCP cult, which is why Jiang Zemin wanted to quickly and completely eliminate Falun Dafa. Falun Dafa was introduced to the human world to rectify and destroy the CCP. It is also the Creator's compassionate salvation for the human world.

If for some reason it is not possible to overthrow the CCP's firewall, then supporting the spread of Falun Dafa in the following ways is also very effective.

The Importance of Supporting Media Platforms Related to Falun Gong

Here I would like to give you an inside look at the traditional Chinese culture Falun Dafa or Falun Gong advocate. Let me tell you a story from Chinese history. In the court of every dynasty in China, there was an official called the royal historian who was responsible for recording history. Their daily job was to record all the big and small things which happened in the emperor's life without being monitored by anyone else. The professional ethics of historians was to write down everything that happened to the emperor exactly as it was in history–this was called the historian spirit.

However, due to China's long history and a large number of dynasties, there were always some emperors who were curious and wanted to know how the historians recorded them. These emperors often had very high-sounding reasons to see what they had done in the past, what they did wrong, and what they did right so they could change it in the future so they could use their own history as a reference. However, all emperors were rejected by the historians, because the historians could not give this permission to any emperor. The historians would tell the emperor directly, "No. Because when the historian knows that the emperor may read what he records, he will feel pressured when he writes. Should this sentence be more obscure, or should it not be recorded at all? So, this is why Your Majesty cannot read it."

In 548 BC, Cui Zhu, a powerful official in Qi state, killed the king of Qi. After killing him, Cui Zhu called the historian Tai Shi Bo. Cui Zhu told him, "Now you just write this: On the 11th day of the fifth month of summer, the king died of malaria." Tai Shi Bo wrote on his bamboo slips, "On the 11th day of the fifth month of summer, Cui Zhu killed his king." Cui Zhu got angry when he saw it, "Didn't I ask you to write that the king died of malaria? Why did you write that I killed him?" So, he killed Tai Shi Bo.

In ancient China, being a historian was like doing business in modern times. The grandfather passed the business to the father, the father passed it to the son, and the son passed it to the grandson. All the descendants of the historian family were engaged in the same profession. Cui Zhu called Tai Shi

Bo's younger brother, Tai Shi Zhong, to be his next historian, but Tai Shi Zhong also wrote those words: "On the 1st day of the fifth month of summer, Cui Zhu killed his king." As a result, Tai Shi Zhong was also killed. Then Tai Shi Bo's third brother Tai Shi Shu also came and wrote: "On the 1st day of the fifth month of summer, Cui Zhu killed his king." Cui Zhu killed Tai Shi Shu.

Finally, their youngest brother, Tai Shi Ji, came and wrote: "On the 1st day of the fifth month of summer, Cui Zhu killed his king." After reading it, Cui Zhu asked, "Don't you know how your three brothers died?" He replied, "I know." Cui Zhu asked again, "Aren't you afraid of death?" Tai Shi Ji replied: "I am afraid. But I am a historian. If I cannot record the true history, it would be my dereliction of duty as a historian. It is better to die than to fail in my duty. You want me to record a false history. Although I am afraid, I would rather die." At that time, Cui Zhu was shocked by his righteousness and threw the bamboo slips at him and said, "Okay, take it and file it." In this way, Tai hi Ji survived. The story is not over yet. When Tai Shi Ji left the palace, he saw another staff member of the History Museum, named Nan Shi, who also came with bamboo slips. Nan Shi expressed, "I am worried that you will be killed by Cui Zhu because you wrote the true history. If that happens, I am afraid that no one will remember that Cui Zhu killed the king on the 1st day of the fifth month of summer. So, I came. I will wait until you die and then I will continue to write true history" Then he questioned Tai Shi Ji, "How did you get out alive? I want to see if what you wrote is true." When Nan Shi saw that the bamboo slips recorded the true history, he left with peace of mind.

This story shows the historians in ancient China had such spirit they would rather lose their lives and pay any price to record history. In ancient China, not only historians, but all walks of life regarded their work as a social responsibility. At that time, every Chinese had to fulfill their family and social responsibilities. But after the communists occupied China and launched the Chinese version of cancel culture, also known as the Cultural Revolution, China's divine culture was replaced by the Marxist cult from the West. People became atheists and only worshipped money and power. This made China, a big country, not only fail to lead the moral improvement of mankind but also become the root cause of all man-made disasters. When the fate of mankind was at stake, the Creator came to the world and reincarnated as a "Chinese" and used his divine power, compassion, and knowledge to guide the world to return to the divine culture. The establishment of Falun Gong and its affiliated media platforms to tell the truth was his specific way to annihilate evil.

Although each of our bodies is created by different gods in their images, which leads to different races, skin colors, and cultures, our souls are all given by the Creator, we all have the same universal values. The Creator created everything in the greater cosmic body, including human beings, so he understands human nature best. This is why he has always emphasized the importance of telling the truth.

The truth is always what the devil and his dictatorial agents on earth fear the most. In May 2024, North Korea conducted a large-scale airdrop of balloons filled with feces and garbage to many parts of the Republic of Korea. As a response, in October of the same year, the South Korean government used drones to drop leaflets to Pyongyang, the capital of North Korea. The content of the leaflets was to inform the North Korean people how many tons of food Kim Jong-un's famous watch could be converted into; how many tons of food Kim Jong-un's daughter's brand-name down jacket could be converted into; and how many years these foods can feed the hungry North Korean people. These very common truths scared Kim Jong-un so much that he blew up all the roads and railways connecting South Korea and North Korea. Kim Jong-un is afraid when South Korea "provokes a military invasion", the North Korean people will rise after learning the truth and attack his regime from both inside and outside.

Falun Gong and its affiliated media platforms not only spread the truth overseas, but also many Falun Gong practitioners make long-distance calls to mainland China to tell the truth to the Chinese people. The CCP's long-term suppression of Falun Gong is also for this reason. It does not want the people in China to know the true face of the CCP's evil. The CCP also worries when it launches a war in the Taiwan Strait, the Chinese people who know the truth will rise and overthrow the CCP regime. If we want to avoid the Third World War, a nuclear war that destroys humanity caused by the CCP's invasion of Taiwan, then the most effective and lowest costing way is to support Falun Gong and its media platforms which tell the truth is the way to go!

There are also some very good media platforms in the United States composed of patriots. The difference of Falun Gong is they have also inherited the spirit of historians in traditional Chinese culture. No force in the world can tempt or threaten them to force them not to tell the truth. To put it simply, Falun Gong practitioners will always measure everything they do with the principles of truthfulness, compassion, and forbearance. Even if Master Li Hongzhi is no longer on the earth, they will stick to this principle. They will never make fake news, even if their lives are threatened.

To put it more extreme, even if someone chopped off the heads of Falun Gong practitioners, the next batch of Falun Gong practitioners working on these platforms would still tell the truth and refuse to create fake news. This is because they are not afraid of death. How do you threaten a group of people who are not afraid of death?

Over the past few years, we have all seen how much damage the fake news media has done to our country and the rest of the world. Having a group of social media platforms that are not tempted by profit, not afraid of death threats, and can insist on telling the truth is crucial to defending our Constitution and preventing uninformed people from being misled by fake news and threatening the future of this country and the world.

The devil's followers have seized almost all the important social platforms to attack our freedom of speech. We should also have our own social platforms to fight against them.

First, support the performance of Shen Yun Performing Arts.

Shen Yun has been telling the truth to the world through art over the years and has played a very important role in letting the world understand the true face of the CCP and what is the true Chinese traditional culture. The director, choreographer, and some actors of Shen Yun Performing Arts are not ordinary people. Many audiences who have seen Shen Yun Performing Arts say Shen Yun Performing Arts performances are the number one show in the world. Many audiences feel when watching Shen Yun's performances, there is a very strong energy field on the scene, and they feel this energy field seems to bring them a kind of spiritual healing.

Second, support the self-media platform GanJingWorld.

Entrepreneurs and God's people who are just and have resources like Elon Musk can only publish important information about your company's products and services to increase the influence and traffic of this platform. This is much more cost-effective and timesaving than buying a huge self-media platform to ensure free speech and avoid censorship. You can make full use of GanJingWorld. This platform is established by people of faith, so your freedom of speech is guaranteed. At the same time, there will be no content on this platform that can negatively affect the growth of children.

Third, subscribe to *The Epoch Times.*

One of the reasons why *The New York Times* attacked Falun Gong and Shen Yun is because *The New York Times* competitor, *The Epoch Times*, is affiliated with Falun Gong. *The Epoch Times* adheres to the principles of truthfulness, compassion, and forbearance, and only reports objective truth. It will never succumb to any pressure or interests to create fake news. Subscribing to *The Epoch Times* will let you know the truth behind major events happening every day in this world. On top of that, readers who want to explore more truths and like my writing style can often see other articles I contribute to *The Epoch Times.*

Fourth, watch NTD News.

NTD is known for its timely, objective, and accurate investigative news reports. As early as early 2020, it investigated and reported that scientists from the Wuhan Institute of Virology collaborated with the CCP's military to combine the world's deadliest coronavirus to create a new variant virus, COVID-19 or CCP Virus. However, due to its limited influence at the time, it did not attract public attention, which indirectly led to countries not taking precautionary measures in advance. NTD Television is one of the best choices to replace traditional mainstream fake news media.

All in all, we need to give the CCP a taste of their own medicine by launching an unrestricted war that includes the destruction of the CCP's finances, aid to the mature Chinese anti-communist revolutionary armed forces, and assistance to the truth-spreading Falun Gong in U.S., China, and the rest of world.

Please remember that morality is the key to your immortality. Your attitude toward the CCP's state terrorism is God's standard for measuring your morality!

03/09/2025
Mark C. Wang

www.ingramcontent.com/pod-product-compliance
Lightning Source LLC
LaVergne TN
LVHW100303210926
848480LV00003B/1006
* 9 7 9 8 9 9 2 8 6 1 7 0 9 *